HOG MANURE MANAGEMENT, THE ENVIRONMENT AND HUMAN HEALTH

HOG MANURE MANAGEMENT, THE ENVIRONMENT AND HUMAN HEALTH

Tiffany T. Y. Guan

University of Manitoba
Winnipeg, Manitoba

and

Richard A. Holley

University of Manitoba
Winnipeg, Manitoba

Kluwer Academic/Plenum Publishers

New York, Boston, Dordrecht, London, Moscow

Library of Congress Cataloging-in-Publication Data

ISBN: 0-306-47807-2

Copyright © 2003 by Kluwer Academic / Plenum Publishers, New York
233 Spring Street, New York, New York 10013

http://www.kluweronline.com

10 9 8 7 6 5 4 3 2 1

A C.I.P. record for this book is available from the Library of Congress.

Permission for books published in Europe: permissions@wkap.nl
Permissions for books published in the United States of America: permissions@wkap.com

Printed in the United States of America.

Preface

This volume provides a current look at how development of intensive livestock production, particularly hogs, has affected human health with respect to zoonotic diseases primarily transmitted by food but also by water, air and occupational activity. While information presented focuses on the development of increasing livestock production in Canada, examples are given and comparisons are made with other countries (Denmark, Taiwan, the Netherlands and the United States) where the levels of livestock production are much more intense and where the industry is more mature. Canada is also searching for solutions to enable handling the growing volume of its livestock waste properly. Lessons learned from the experience of those who have gone before are invaluable and are drawn together in this volume to serve as useful guidance for others in plotting the courses of action possible to avoid serious environmental setbacks and negative human health effects through foodborne illness.

A significant portion of the text is devoted to a discussion of enteric illness in humans caused by zoonotic pathogens. The second chapter deals with survival of pathogens (which cause foodborne illness) in manure environments. An evaluation of the human health hazard likely to occur from the use of manure as fertilizer is important because of the recent trend toward an increase in foodborne illness from the consumption of minimally processed fruits and vegetables that may have been fertilized with animal-derived organic materials. Pathogen survival in manure, soil, and water is described in as much detail as the literature will currently permit. The relationship between intense hog production and human enteric illness is examined in the third chapter with data from all major hog producing countries of the world (Denmark, the Netherlands, Taiwan and the United States). Another chapter deals with the reliability of sampling and methods used for the detection of zoonotic bacteria in agricultural samples. Organisms discussed include the major bacterial causes of foodborne illnesses, protozoans (cryptosporidia, giardia) and viruses. A chapter is provided that deals with methods used for manure handling and treatment which is prefaced by explanation of primary, secondary, and tertiary treatment of wastewaters. Information is provided on how industry practice in a variety of countries differs and offers an explanation of why there are differences. A final chapter examines legislative initiatives and associated policies driving change in the industry and identifies important factors which must be considered in the development of regional best practices for manure handling to minimize risks to health and the environment.

The purpose of this volume is to establish the current state of knowledge regarding the influence of intensive hog rearing practices upon zoonotic pathogen presence and movement in the environment. It examines in detail evidence linking these intensive operations to the health status of individuals living in countries having high levels of swine production. Data used have been collected from the scientific literature, from professional associations and government agencies worldwide as well as through industry reports. This volume contains an extensive and fully annotated bibliography containing almost 400 references. It summarizes and discusses information available from experience in countries where intensive swine rearing is a more mature industry to facilitate the orderly growth of the industry elsewhere, where it is less mature, and minimize the impact of its growth upon the environment and human health. Linkages between pork consumption (but not swine-manure handling) in Denmark and the Netherlands and gastroenteritis caused by *Yersinia* or *Salmonella* serotypes from hogs is established and discussed. The reliability of methods used for the recovery of zoonotic bacteria, viruses and protozoans from environmental samples is evaluated. Gaps in knowledge are identified and areas requiring additional work are pointed out.

To assist in the development and application of best practices for manure handling, the ability of bacterial pathogens, viruses and protozoans (likely to be present in manure) to survive in manure, compost, soil and water is examined in detail. Data available show that pathogens of concern can be eliminated by holding manure at $\geq 25°C$ for 3 months. Predominant manure handling practices are examined at an international level and factors influencing choice of practices used regionally are discussed. This information can be used, in turn, as background by decision makers in planning for industry growth. Regulatory response to intensive hog rearing and the impact this has had upon industry development is provided to serve as future guidance for both the industry and legislators. Priorities for future research are identified. It is intended that the data accumulated be used to validate best manure management practices at established and predicted production thresholds.

Richard Holley
Winnipeg, Manitoba, March 2003

Acknowledgments

Gary Graumann contributed to the final version of the index. This book was supported by funds from the Manitoba Livestock Manure Management Initiative and the Manitoba Rural Adaptation Council.

Contents

CHAPTER 1

Sources of Enteric Disease in Canada

SUMMARY

The public is exposed to the risk of enteric infection via foodborne, waterborne, and airborne sources, or via direct contact by infected persons or animals. Foodborne transmission of enteric disease is the most well-known and extensively studied area. Among all foods, meat and poultry products are the largest contributor of foodborne disease outbreaks. Consumption of these products has been linked to enteric infection caused by several important human pathogens such as *Escherichia coli* O157:H7, *Salmonella enterica* Typhimurium, *S.* Enteritidis, *Campylobacter* species, and viruses. Food animals carry some of these human pathogens with no clinical symptoms. For example, *E. coli* O157:H7 carried by young cattle, *Salmonella* and *Campylobacter* by poultry. Carriage of human pathogens in food animals has serious health implications because it means the farm is a significant reservoir of foodborne illness agents. The ecology of the most frequently occurring human pathogens in the farm environment is reviewed here. Other important vehicles of foodborne gastroenteritis include fruit and vegetables, seafood, and dairy products. Fruit and vegetables, as raw agricultural commodities, are vulnerable to microbial contamination during production, and minimal processing further magnifies the contamination level. The opportunity for fresh produce contamination at production is discussed. Seafood-associated enteric infection is usually the result of environmental contamination and eating raw or undercooked seafood. The frequency of environmental contamination of seafood is examined. Enteric infection due to dairy foods often implicated raw milk, improper pasteurization, and post pasteurization contamination as sources of the problem. Some of the dairy product-associated outbreaks are addressed. Waterborne transmission is a second common route for spreading enteric disease. However when it happens, it usually results in a large number of people being affected. Drinking and recreational waters are two main vehicles. Important waterborne disease agents such as *Cryptosporidium*, *Giardia*, *Shigella*, and *E. coli* O157:H7 are reviewed. Airborne transmission of enteric infection is the least studied area. Airborne microbial contamination and the closely related occupational hazards are acknowledged. Some of the enteric disease agents can be spread through direct contact either by infected animals or humans themselves. Among agents spread by animal contact are verocytotoxigenic *E. coli*, predominantly *E. coli* O157:H7, *Salmonella* species including *S.* Typhimurium DT104, and *Cryptosporidium*. Enteric infection by personal contact has been documented for Norwalk virus, hepatitis A, *Shigella* species, *E. coli* O157:H7, and *Cryptosporidium*. The role of direct contact in disease transmission is addressed.

INTRODUCTION

Gastroenteritis is a common human illness. Each year millions of children worldwide die from diarrheal diseases (WHO 1999). In the United States, enteric disease ranks second in prevalence to respiratory disease (Lindsay 1997). Humans can contract gastrointestinal diseases via food, water, air, and by direct contact with ill persons. In the U.S., the percentage of diarrheal illness due to foodborne transmission is 36% (Mead et al. 1999). This should not seem surprisingly low because foodborne pathogens can also infect humans via nonfood routes. For example, although foods of animal origin are the major source of *E. coli* O157:H7 infection, transmission through fresh produce, water, and direct contact with infected people and animals has been increasingly recognized. In Ontario, more than half of all outbreaks of enteric disease reported in 1993–1996 were transmitted from person to person (Health Canada 1998). Therefore, the reported rates of foodborne disease from some surveillance data do not represent foodborne sources exclusively (CDC 2001).

Most enteric diseases result in clinical conditions associated with acute diarrhea, vomiting, fever, abdominal cramps, or other gastrointestinal manifestations such as dysentery (Lindsay 1997). However, some enteric pathogens can cause more severe health consequences including chronic sequelae such as kidney failure, renal disease, brain and nerve disorders, and even death (Lindsay 1997). Despite advances in modern food and medical technologies, enteric diseases remain a major cause of morbidity and mortality. Identifying and controlling of enteric disease agents, therefore continue to be challenges for health officials, scientists, and researchers. The epidemiology of infectious diseases including enteric disease is rapidly changing. Factors contributing to this observation include changes in human demographic characteristics, human behavior and consumption habits, consolidation of industries and new technology, globalization of food trade, international travel, microbial evolution, and breakdown of public health infrastructure (Altekruse et al. 1997).

Human demographic changes have led to increased numbers of susceptible persons, such as the immunocompromised (e.g. AIDS patients), elderly (aging population as a whole), and chronic disease patients (e.g. expanded life expectancy of cancer patients due to advanced medical technologies) (Altekruse et al. 1997). Historically, enteric diseases pose the greatest threat to young children, pregnant women, and the elderly (WHO 1999). Changes in lifestyle and eating habits also have had an influence on the occurrence of enteric diseases. Dining out at restaurants has increased during recent decades. Outbreaks outside the home accounted for almost 80% of reported outbreaks in the U.S. in the 1990s (Altekruse et al. 1997). Increased consumption of fresh fruits and vegetables in recent years may have contributed to the increased incidence of disease outbreaks associated with fresh produce. Consumption of high-risk foods, such as pink hamburgers, runny eggs and raw shellfish, and unsanitary food handling practices at home also add to the risk of acquiring enteric diseases (Shiferaw et al. 2000). Consolidation of the agriculture and food industries combined with modern transportation technology has allowed greater geographic distribution of products from large centralized food processors (Altekruse et al. 1997). This mass distribution of foods has led to more frequent large and widespread outbreaks of enteric disease. Many outbreaks of disease have involved more than one country, and some more than one continent (WHO 1999). Globalization of food trade, therefore, presents a transnational challenge to food safety authorities because food contaminated in one country can result in outbreaks of disease in another. International travel has increased dramatically during the 20th century. Travelers may become infected with pathogens uncommon in their nation of residence, and carry pathogens home to infect nontravelers (Altekruse et al. 1997). The consequences include complications in diagnosis and treatment especially when the symptoms begin only after travelers return home. Microbial evolution is a major force influencing emerging diseases. Microorganisms change and adapt in unfavorable environments. The use of therapeutic antimicrobial agents, in both human and animal populations, creates a selective pressure that favours survival of resistant bacterial strains (Altekruse et al. 1997). The widespread use of sub-therapeutic levels of antimicrobial agents for promotion of growth in livestock may also select for resistant bacteria that are harmful to humans. In addition, rapid mutation of some human pathogens such as *Escherichia* and *Salmonella* adds to the challenge of controlling causal pathogens. Further, as many public health agencies operate with extremely limited resources, investigations examining the occurrence of illnesses may be delayed. This interruption in public health infrastructure increases the potential for underreporting diseases.

For the above reasons, many experts believe the risk for foodborne enteric illness is increasing (WHO 1999; Collins 1997). Others find it difficult to conclude whether or not foodborne illness is on the increase. This lack of consensus is due, in part, to technical improvements in the detection methods being used and to the likelihood that apparent increases in incidents are in some respect artifacts of contemporary focus (Doores 1999). Reported incidents of *E. coli* O157:H7 infection, for example, have increased ever since the organism was recognized as a human pathogen two decades ago. This may have been the result of increased awareness by health authorities and in laboratories.

Adding to the complex epidemiology of enteric diseases is that the source of illness in most outbreaks is usually unknown (Todd et al. 2000; Tauxe et al. 1997). There are a number of reasons why it is sometimes impossible to identify a causal agent (Doores 1999; Tauxe et al. 1997): (1) the illness is not of infectious origin; (2) a sample is not obtained at the appropriate time during the patient's illness; (3) the current detection techniques do not include a test for a specific causal pathogen, or the pathogen is a virus that is not detected by current methods; (4) the causal organism is a little understood newly emergent pathogen, or an unknown pathogen; and, (5) if a multi-ingredient food is implicated in an illness, it is not always possible to identify which ingredient is the actual source of the contaminating pathogen.

Addressing enteric diseases will require not only more sensitive, rapid surveillance and enhanced methods of laboratory identification, but also effective prevention and control. The general strategy of prevention is to fully understand the mechanisms by which contamination and disease transmission occur so as to allow interruption of the processes. There are both research and education elements involved in developing requisite understanding. The prevention of enteric diseases via foodborne routes will depend on careful food production, handling of raw products, and preparation of finished foods. Whereas the prevention of diseases via waterborne routes will rely on careful monitoring of municipal water supplies, well water sources, and recreational waters (Altekruse et al. 1997). Technologies are available to prevent most foodborne and waterborne illnesses. They include good food sanitation and hygiene practices plus techniques such as pasteurization, refrigeration, retort canning and water chlorination (Altekruse et al. 1997). The use of Hazard Analysis Critical Control Points (HACCP) programs also helps to identify and control hazards and subsequently prevent foodborne diseases. Other technologies that are worth our consideration are irradiation, chlorination of drinking water for food animals, composting of manure used as fertilizers, chlorination of irrigation water, and other technologies that reduce the number of potential pathogens present in agricultural commodities (Altekruse et al. 1997).

To date, at the beginning of the 21st century, enteric disease remains a major threat to public health. Although a significant amount of research has been focused on the genetics, physiology, and virulence of human enteric pathogens, it is equally important to understand how these microbes, humans, and animals interact in the environment. A better understanding of the mechanisms by which these pathogens persist in animal reservoirs and environments is critical to successful long-term prevention. Together, this information will provide an intellectual and technological foundation upon which effective control programs and disease prevention strategies can be built. This chapter will review the existing scientific knowledge of human enteric disease, more specifically its sources in the environment and the avenues through which pathogens cause disease in the human population.

FOODBORNE TRANSMISSION

Foodborne transmission is probably the most common route for transmission of enteric diseases. Foods of animal origin account for the largest percentage of outbreaks and sporadic cases in Canada (Todd et al. 2000). Other important food vehicles include marine foods, dairy products, and fruit and vegetables. In this section, sources of contamination associated with each category of food based on previous outbreaks, results of outbreak investigations, and the possibilities of other transmission routes particularly at the farm level will be discussed.

Meat and Poultry

Meat and Meat Products

Some enteric pathogens are more likely to be associated with a specific kind of animal product. This is because many human pathogens have an animal reservoir from which they spread to humans. These pathogens, called foodborne zoonoses, do not often cause illness in the infected host animal (Tauxe 1997). Consumption of bovine foods, for example, is frequently associated with verocytotoxin-producing *Escherichia coli* (VTEC) O157:H7 infection. Beef and dairy cattle are therefore believed to be the primary reservoir and the dairy farm is considered a risk setting for transmission of the pathogen. In the United States, an estimated 73,000 cases occur annually with an estimated 61 fatal cases (CDC 2000). The pathogen is one of 9 pathogens under surveillance by the Foodborne Diseases Active Surveillance Network (FoodNet), an Emerging Infections Program of the Centers for Disease Control and Prevention (CDC). It is estimated that 75% of hemorrhagic colitis and hemolytic uremic syndrome cases in Canada are attributed to infection with *E. coli* O157:H7 (Health Canada 1999). Studies in Ontario, Canada, have found *E. coli* O157:H7 contamination in cattle on farms and abattoirs to vary from <0.3% to 1.5% (Cassin et al. 1998). Earlier studies by Wells et al. (1991) found the prevalence of *E. coli* O157:H7 shedding in calves and heifers to be 2.3 and 3%, respectively, in contrast to a 0.15% incidence of fecal shedding in adult cattle. A recent Canadian study by Van Donkersgoed et al. (1999) reported isolation of *E. coli* O157:H7 from 7.5% of fecal samples collected from cattle at slaughter.

E. coli O157:H7 colonizes the gastrointestinal (GI) tract of cattle and the infected cattle may appear perfectly normal. This is significant because there is no way to visually recognize the contaminated animals and prevent them from going to slaughter. It is possible that the feces of animals carrying *E. coli* O157:H7 may periodically test negative. Therefore the traditional means of controlling infectious agents, such as eradication or removal of carrier animals, are not feasible for this organism. Since shedding of the organism from cattle is only transient, routine testing may not identify infected animals. Since the prevalence of the organism is generally low in cattle herds, it is difficult with current sampling methods to detect the presence of *E. coli* O157:H7. This is of some significance when one considers the low infectious dose of the organism (it takes 10 cells or less to trigger disease in humans). *E. coli* O157:H7 seems to be a common resident in ruminants, such as cattle, sheep, deer, and goats. An *E. coli* O157:H7 infection was associated with consumption of jerky prepared from deer meat (Keene et al. 1997). The pathogen has also been isolated from pigs (Chapman et al. 1997). However, to date, there has been no direct evidence of human infections originating from pigs in North America (DesRosiers et al. 2001). Although an increasing variety of food sources and other vehicles for *E. coli* O157:H7 transmission have been documented in recent years, direct or indirect exposure to bovine or human feces explains almost all cases where a source is identified. Research has been undertaken to elucidate the ecology of *E. coli* O157:H7, with particular emphasis on the identification and control of sources of the pathogen in the farm environment. One research study is being conducted to test a vaccine for cattle in order to reduce the level of colonization of the host by the pathogen and reduce shedding (Potter et al. 2000).

Over the past decade, several research groups have defined many of the major features of the epidemiology of *E. coli* O157:H7 on farms. (1) It is nearly ubiquitous in cattle populations; the pathogen can be found on virtually all farms, at least intermittently. (2) Its occurrence is strongly seasonal with peak prevalence in summer and early fall. (3) It shows transient residence in the GI tract of animals and is not associated with clinical disease. (4) There is higher prevalence in young animals. (5) There is a lack of host specificity; the pathogen has been isolated from humans, cattle, sheep, dogs, cats, birds, horses, flies, etc. (6) There is temporal clustering in cattle populations; most fecal shedding is confined to sharp bursts in a high percentage of animals separated by long periods of very low prevalence. (7) There is a complex epidemiology at the molecular level; several pulsed field gel electrophoresis (PFGE) subtypes (having different DNA "fingerprints") often exist on a farm simultaneously with periodic additions and turnovers, even on farms that do not receive animals from outside. (8) The organism is capable of long distance transmission not associated with animal movements; the same PFGE subtypes are found on closed farms (absence of any obvious animal movement between them) separated by hundreds of kilometers (Hancock et al. 2001).

Several researchers have been trying to provide explanations for the above unique and complex features of *E. coli* O157:H7. Subtypes of the organism can persist on cattle farms for years, and common subtypes are often found in environmental niches and in other species of animals, thus supporting a conclusion that cattle farms represent a reservoir, but it is not completely clear whether cattle themselves are the reservoir (Hancock et al. 2001). The seasonality of the pathogen in cattle also correlates with seasonal human infection and contamination of retail meat. Previous speculation attributed the summer increase of human *E. coli* O157:H7-associated disease to seasonal consumption patterns (e.g. barbeque cooking), but it did not explain the seasonal contamination of retail meat. Jones (1999) interpreted the seasonality as a reflection of the movement of cattle from winter housing to summer grazing in the spring, and in the autumn to a return back to winter housing, which was accompanied by changes in diet and water source which induced stress. Bunk feeds (feed mixes directly consumed by cattle) collected from feeding troughs and cattle watering troughs are commonly positive for *E. coli* O157:H7; thus feed and water likely represent the most common means of infection on farms. Environmental replication of

E. coli in feeds and in sediments of watering troughs does occur and this may account for the higher level of fecal shedding of *E. coli* O157:H7 in the warm months (Hancock et al. 2001). The pathogen has also been found to persist and remain infective for at least 6 months in water trough sediments; it is hypothesized that this may be an important environmental niche where the pathogen survives during periods when it cannot be detected in cattle, especially during cold months. Therefore, certain farm management practices, especially those related to maintenance and multiplication of the infectious agent in feed and water, may provide a practical means to substantially reduce the prevalence of *E. coli* O157:H7 on farms. Further, the pathogen is able to remain viable in soil for greater than 4 months (Jones 1999). This has serious implications for the land-based disposal of organic wastes such as cattle manure, cattle slurry and abattoir waste. The presence of *E. coli O157:H7* in wild bird populations, mainly gulls, has prompted questions about the cause of this contamination outside the farm. It is speculated that birds become contaminated after feeding on pastureland following the application of farm slurries and sewage sludge (Wallace et al. 1997 cited by Jones 1999).

Undercooked ground beef, such as the hamburger patty, is a common vehicle implicated in *E. coli* O157:H7 infection. Recently, two outbreaks in Canada were associated with ready to eat foods such as dry fermented salami. In the spring of 1998, an *E. coli* O157:H7 outbreak was traced to a naturally fermented Genoa Salami product manufactured by a registered establishment in Ontario (Health Canada 2000a). Again in November 1999, another *E. coli* O157:H7 outbreak in British Columbia was traced to a similar type of raw fermented Hungarian-style sausage (Health Canada 2000a). Research has found that methods used to manufacture fermented sausage do not control or eliminate this acid tolerant pathogen from the finished product. These observations and the serious consequences of the outbreaks have prompted Health Canada to impose regulations on the sausage industry for controlling this organism in the manufacture of dry and semidry fermented sausages (Health Canada 2000a). Manufacturers are required to follow one of several interventions which include high temperature cooking (62.8°C for 4 min); a combination manufacturing process that achieves a 5 $\log_{10}$ (5D) reduction in numbers; end product testing with the lot held pending receipt of results; implementation of a HACCP program that includes raw material testing and a process that achieves 2D reduction; or, any alternative manufacturing process that is scientifically validated to achieve the required *E. coli* O157:H7 reduction.

In addition to *E. coli* O157:H7 infection, salami has also been associated with outbreaks of *Yersinia* and *Salmonella* (Health Canada 2000b; Pontella et al. 1998). Unlike *E. coli* O157:H7, *Yersinia enterocolitica* is widely recognized as a cause of disease in animals. Not all strains are pathogenic to man. Surveys in some countries have shown that pig's tonsils frequently carry pathogenic serotypes. Others have found relatively high numbers in the caecal contents of infected pigs. The pathogen is of particular concern in food because it is able to grow at refrigeration temperatures (Mackey 1989). Studies of the association between consumption of sausage products and yersiniosis linked *Yersinia*

Table 1.1. Incidence per 100,000 population of diagnosed foodborne
illnesses at the five original sites (1996–2000) and for all eight sites (2000),
by year and pathogen—Foodborne Diseases Active Surveillance Network,
United States.[a]

| Pathogen | Original five sites | | | | | All sites |
	1996	1997	1998	1999	2000	2000
Campylobacter spp.	23.5	25.2	21.4	17.5	20.1	15.7
Cryptosporidium parvum	NR[b]	3.7	2.9	1.8	2.4	1.5
Cyclospora cayetanensis	NR	0.4	0.1	0.1	0.1	0.1
Escherichia coli O157:H7	2.7	2.3	2.8	2.1	2.9	2.1
Listeria monocytogenes	0.5	0.5	0.6	0.5	0.4	0.3
Salmonella spp.	14.5	13.6	12.3	13.6	12.0	14.4
Shigella spp.	8.9	7.5	8.5	5.0	11.6	7.9
Vibrio parahaemolyticus	0.2	0.3	0.3	0.2	0.3	0.2
Yersinia enterocolitica	1.0	0.9	1.0	0.8	0.5	0.4

[a]Reprinted from CDC (2001a) with permission.
[b]NR—not reported.

species to porcine reservoirs. Undercooked pork products have repeatedly been
associated with yersiniosis (Tauxe et al. 1987; Ostroff et al. 1994; Satterthwaite
et al. 1999), which results in gastroenteritis and appendicitis-like symptoms.
More serious cases can lead to polyarthritis, septicemia, and meningitis (Health
Canada 2000b). These serious health consequences, along with the high incidence of yersiniosis in British Columbia (28 per 100,000 persons in 1998) make
this disease a significant public health problem in the province. Research is
needed to review the parameter limits set for the intrinsic factors (pH and water
activity, Aw) of dry fermented meat products to control the pathogen. In addition, research is also needed to identify risk factors associated with *Yersinia*
infection and consumption of pork products.

Outbreaks of *Salmonella* species associated with meat and meat products are frequently subjects of contemporary news reports. As with *Yersinia*,
Salmonella have been associated with illness among many animals, including
cattle. *Salmonella* live in the intestinal tracts of various animal species including cattle and they represent a major reservoir for human disease. In humans,
this enteropathogen is among the most commonly reported and costly causes
of foodborne disease. It is estimated by CDC that 95% of human salmonellosis
in the U.S. occur through foodborne transmission (Mead et al. 1999). FoodNet
data from the U.S. reported there was an incidence rate of 14.4 cases per 100,000
persons during 2000 at a total of eight FoodNet surveillance sites (Table 1.1).

Can *Salmonella* reduction be achieved on cattle farms? *Salmonella* species
are common in outflows from human sewage treatment plants and may therefore
contaminate surface water and animals consuming it downstream (Wells et al.
2001). Animal and plant protein feed may also be contaminated with *Salmonella*
and contribute to clinical disease in cattle. Wild birds and rodents may also
be a source of *Salmonella* contamination. These multiple sources complicate

development of on-farm control strategies, since an overall prevention strategy needs to consider all off-farm inputs (such as livestock drinking water, feed sources, sources of water used to irrigate pasture and crops for livestock feed) in addition to on-farm control strategies. Despite these environmental complexities, research from other species of livestock and poultry indicates that herd management can reduce shedding of *Salmonella* (Wells et al. 2001).

On hog farms, pigs can become infected with *Salmonella* directly via consumption of contaminated feed, contact with another infected pig, or directly through a contaminated environment, such as surfaces, equipment, or nonporcine animals including humans (Hume, 2001). It is believed that the lairage where pigs are held from 2–8 h before they are slaughtered can act as an important source of *Salmonella* infections of animals in healthy herds (Swanenburg et al. 2001). The efficiency of cleaning and disinfection of the lairage must be ensured to eliminate this risk. It was found that long-term persistence of *S*. Typhimurium on hog farm premises led to recurrent infections (Baloda et al. 2001). One of the explanations offered is 're-inoculation' of the pathogen from the herd environment. When *Salmonella*-contaminated slurry was disposed of on agricultural soil, a common waste disposal practice, the pathogen could be isolated up to 14 days after being spread (Baloda et al. 2001). This creates potentially high risks for transmission of the pathogen in the environment, animals, and humans. In the U.S., 8.7% of swine carcasses were positive for *Salmonella* before implementation of the Clinton Administration's science-based meat and poultry inspection system (FSIS 1996). In 2000, the prevalence was reduced to 4.1% in large plants, 7.7% in small plants, and 7.2% in very small plants (FSIS 2001). Hazard analysis critical control points (HACCP)-based inspection systems are credited as being effective in reducing prevalence of *Salmonella*-contaminated raw meat and poultry.

The gut of healthy animals harbours other important human enteric pathogens such as *Campylobacter jejuni*. In the past many years, *Campylobacter* species consistently caused the largest number of foodborne incidences where the etiological agent was identified (Table 1.1). *Campylobacter* spp. colonize the gastrointestinal tracts of a wide range of wild and domestic animals, including cattle, hogs, and chickens without clinical effects (Kramer et al. 2000). Their presence on the carcass is therefore to be expected. However, the organisms are sensitive to stresses such as drying, chilling, and exposure to oxygen (Mackey 1989). Mishandling and cross-contamination of red meat and its products are frequently the primary risk factor in transmission to humans. Although red meat and offal are potential sources of this type of enteritis, outbreaks associated with consumption of poultry or poultry products are more extensively documented (see next section).

Listeria monocytogenes is also found among the natural gut microflora of healthy animals and is ubiquitous in farm environments. It is an opportunistic pathogen generally affecting those who are in some way immunocompromised, but this is not always the case, healthy adults can become ill. The especially at risk groups include pregnant women and neonates, individuals receiving immunosuppressive drugs, and those suffering from alcoholism (Mackey 1989).

The organism has frequently been isolated from red meats and can be shed in large numbers in the milk of cows suffering from *Listeria* mastitis.

Staphylococcus aureus is an enterotoxin-producing pathogen carried by animals. However it does not compete well and does not grow on properly refrigerated meat. There is a requirement for more than 10^6 organisms per gram of food in order to produce enterotoxin. Humans are more likely to be the primary source of the organism associated with contamination (CDC 1997a). Ham is the most common reported vehicle of transmission in staphylococcal food poisoning. The salt content of precooked, packaged hams is high, often as high as 3.5%, which provides an ideal growth medium for the pathogen when the temperature is favourable. Staphylococcal poisoning is usually a result of cross-contamination from infected food handlers and equipment surfaces. Better availability and use of refrigeration systems have reduced the frequency of outbreaks caused by this organism.

Eating wild game has frequently been associated with trichinosis or trichinellosis. Trichinellosis is a widespread helminthic zoonosis endemic in Northern Canada where the estimated incidence rate in the Aboriginal populations is 11 cases per 100,000 (MacLean et al. 1989). Infected polar bear and walrus meat are the most frequent sources of human trichinellosis in the Canadian Arctic although there are other potential carnivore sources including red or arctic foxes, wolves, and wild boar (Health Canada 2001a). One recent outbreak of trichinellosis in the U.S. in 1995 was associated with eating cougar jerky (CDC 1996a). *Trichinella* species are found in virtually all warm-blooded animals. Freezing can kill most species of *Trichinella* including *T. spiralis* except for *T. nativa* and *Trichinella T6* (CDC 1996a). To ensure that all *Trichinella* are destroyed meat should be thoroughly cooked ($\geq 71.1°C$ internal). Undercooked pork has long been known to be a risk factor for acquiring trichinellosis. Domestic swine-associated cases have decreased in recent years due to laws prohibiting feeding offal to hogs, increased freezing of meat, and the practice of thoroughly cooking pork (CDC 1996a).

Poultry and Poultry Products

Salmonellosis and campylobacteriosis are enteric diseases of primary concern associated with poultry and poultry products (Bryan & Doyle 1995). Estimates place the annual incidence of human salmonellosis and campylobacteriosis in the U.S. between one to four million for each respectively, resulting in at least 500 deaths (CDC 1996b). Annual cost estimates of poultry-associated cases of salmonellosis and campylobacteriosis in the U.S. range from $64 million to $114.5 million and $362 million and $699 million, respectively. The need for cost-effective solutions for these poultry-borne human disease problems is apparent.

As with meat, poultry are first contaminated from a variety of sources on farms and then contaminants are spread during transportation, processing, handling in markets and kitchens. Similarly, the risk of poultry-borne enteric diseases is influenced by (a) presence of disease causing agents in fowl and opportunity for spread within flocks and during processing, (b) propagation on farms and within processing plant and kitchen environments, and (c) survival of these pathogens on farms and during processing and cooking (Bryan & Doyle 1995). Intestinal colonization and contamination of body parts including feathers of fowl occur on farms, and are favoured by intensive rearing. Further, these pathogens are transferred between birds during transportation, between carcasses and parts during processing, and therefore are commonly found on poultry and its products after processing. By comparison, more poultry products are contaminated with *Campylobacter jejuni* than with *Salmonella* (Bryan & Doyle 1995). In fact, chicken is much more frequently contaminated with *Campylobacter* than red meats are, and the level of contamination is often high (Pearson et al. 1993).

Chickens, turkeys, and ducks are all potential intestinal carriers of *C. jejuni*, usually with no associated symptoms. The infection differs from that caused by *Salmonella* in that the organism does not normally become established in the chicks before the age of 2 weeks. After that, high numbers may be found in the caecum (Mackey 1989). Although *Salmonella* often are present in the intestinal tract of mammals and birds, they are not a normal part of the intestinal microbiota. They are readily acquired from feed and environmental sources (Bryan & Doyle 1995). The host-adapted *Salmonella enterica* Pullorum, *S. enterica* Gallinarum, and, to a lesser extent, *S. enterica* Enteritidis cause clinical illness in the birds but other strains tend to only colonize the caecum where they may be excreted for varying periods with no obvious symptoms (Mackey 1989).

Eating undercooked eggs and poultry is frequently associated with salmonellosis. Undercooked egg is a common source of *Salmonella enterica* Enteritidis. The pathogen emerged as a major egg-associated pathogen in the late 20th century. Epidemiological data from several countries indicate that *S.* Enteritidis filled the ecologic niche vacated by eradication of *S.* Gallinarum from poultry. Surveys in Germany (Rabsch et al. 2000) demonstrated that the number of human *S.* Enteritidis infections was inversely related to the prevalence of *S.* Gallinarum in poultry. It is believed *S.* Gallinarum competitively excluded *S.* Enteritidis from poultry flocks early in the 20th century. Today, *S.* Enteritidis remains one of the most commonly reported serotypes of *Salmonella*.

Freshly laid eggs rarely contain microorganisms, although *Salmonella* Enteritidis can contaminate yolk via the transovarian route from infected hens (Mackey 1989). Thereafter, the eggshell can be contaminated from chicken feces and *Salmonella* may penetrate through shells and membranes during cooling. The consequences include contamination of chicks from the shells during or after hatching. In contrast to *Salmonella*, eggs derived from hens infected with *C. jejuni* do not yield this organism in yolk, albumen, or shell surface. Therefore *C. jejuni* is not likely to be transmitted by eggs.

Due to the almost ubiquitous distribution of *Salmonella* in the animal environment, fecal material residue, dust, and fluff in the hatchery environment may be important sources of the pathogen that contaminate young poults. Several studies have shown a decrease in prevalence of fecal shedding of *Salmonella* as birds age. Chicks and poults are readily infected with *Salmonella* if the pathogens are present in feed or environment. They may shed as many as 10^8 cells per gram of feces (Bryan & Doyle 1995; Mackey 1989). As birds mature (even though *Salmonella* remains in litter, soil and on fecally contaminated surfaces), they tend to shed fewer organisms and a smaller percentage of the flock remains infected, unless the birds continually receive contaminated feed. Poultry feed is a frequent source of contamination of fowl and farm environments. Although meat and bone, feather and fish meals are treated to kill *Salmonella*, microbes from raw materials often re-contaminate the finished products. Feed however, is not a source of *C. jejuni* due to its low moisture content which causes *C. jejuni* to die off.

In the environment, *Salmonella* and some other enteropathogens can survive in poultry litter and soil from a few days to several weeks. Because fowl peck the ground, pathogens shed in feces and present in litter and soil are ingested. Wild animals such as birds, reptiles, insects, and vermin that reach poultry farms also serve as reservoirs of enteric pathogens (Bryan & Doyle 1995). *Salmonella* and *Campylobacter* have been isolated from drinking water for fowl. Bacterial pathogens can contaminate, survive, and multiply in water if organic matter such as feed and fecal matter is present. The environment in and near poultry rearing houses is a source of *Campylobacter jejuni* for young chicks. The pathogen can also be introduced on the footwear and clothing of farm workers. *Campylobacter* has been found in the air, litter, and drinking water containers in poultry rearing and finishing houses (Pearson et al. 1993).

Intensive rearing of fowl (especially young chicks), where thousands of birds are kept together, is very conducive to the spread of *Salmonella*, and possibly other enteropathogens (Bryan & Doyle 1995; Mackey 1989). Under these conditions, an infected bird, a contaminated lot of feed, or other sources of contamination in the environment can easily spread pathogens to many birds. Transmission of pathogens continues after birds leave the farm, before arriving at the slaughterhouse. A *Salmonella* outbreak traced back to contamination of a transport truck has been documented. Since *Salmonella* and *Campylobacter* species are frequently present in feces, they are transferred to feathers and the skin of birds during transit. Stresses caused by capturing, transporting, crowding, holding in crates before slaughter, and inclement weather increase contamination among fowl. The longer birds are held in shipment crates, the greater is the percentage of positive birds contaminated by pathogens.

Although processing may take place in modern sanitary plants, *Salmonella* and *Campylobacter* are still present on poultry throughout processing operations (Bryan & Doyle 1995). Scalding, defeathering, evisceration and giblet operations are the major points of transfer of pathogens. Often there are greater numbers of processed carcasses and parts contaminated than there are infected live animals coming to slaughter. Unless all poultry is properly heat treated and handled before consumption, additional prevention and control procedures are needed to reduce the risk of poultry-borne enteric disease.

At the consumer level, factors associated with *Salmonella* outbreaks usually include improper thawing of frozen poultry products, use of raw eggs, and undercooking of both poultry and eggs. Preventative action can be taken at final preparation, but this will require education of millions of food handlers and homemakers. To date, prevention strategies have relied on pathogen control at slaughter, refrigeration of cooked meat products, and use of preservatives. Nonetheless, there is an immediate need for a cost-effective approach to reduce the prevalence of *Salmonella* and *Campylobacter* on poultry at the farm level. Farm practices including use of fresh manure as fertilizers for vegetable fields, use of non-potable drinking water for food animals, and inadequate management practices used with high densities of animals in barns and feed lots should be evaluated to assess their potential for transmission of these pathogens. All the above activities can create routes for contamination of produce, animals and their environments.

Food irradiation is an important option for reducing the incidence of food-borne illness caused by microbial pathogens that may contaminate raw meat or poultry. Food manufacturers in the U.S. are currently allowed to irradiate raw meat and poultry as well as several other food products. Irradiation technology employs use of gamma rays, X-rays, or electron beams. The maximum permitted radiation dose for meat (4.5 to 7.0 kGy) and poultry (3.0 kGy) is sufficient to inactivate at least 99.9% of common foodborne pathogens such as *Salmonella* and *E. coli* O157:H7 (Frenzen et al. 2001). The safety of irradiated food has been established by extensive research. Therefore, consumers who substitute irradiated raw meat and poultry for non-irradiated products can reduce their risk of foodborne illness, and those at increased risk should experience the greatest health benefits. However, the demand for irradiated meat and poultry has been low, despite U.S. federal government approval.

Seafood

Most seafood is safe; however, as with all foods it carries some risk. Seafood is the second most commonly reported vehicle implicated in foodborne enteric disease in Canada (Todd et al. 2000). Most risks due to seafood consumption originate from environmental contamination. Pathogens implicated in seafood-borne illness are classified into 4 major categories: 1. those naturally present in marine or freshwater environments; 2. those associated with faecal pollution of the environment; 3. those associated with processing and preparation operations; and, 4. those of unknown etiology (Table 1.2). As in other food categories, the largest number of reported cases have unknown etiologies.

Most seafood-associated illnesses are reported from consumption of raw bivalve molluscan shellfish (Ahmed 1992). Molluscan shellfish concentrate particulate matter, including infectious bacteria, from the water that passes through their filtering system. These concentrates eventually pass to the digestive system of the mollusc. The risk associated with eating shellfish is mainly due to *Vibrio* infection and most incidents appear to occur in coastal areas where shellfish consumption is high. Several outbreaks of *Vibrio parahaemolyticus* infections in North America have been associated with eating raw or undercooked oysters

Table 1.2. Human pathogens associated with seafood and aquaculture.[a]

Origin of pathogens	Pathogens
Indigenous, naturally present	*Vibrio* spp. including *V. cholerae* non-O1, *V. parahaemolyticus*; *V. vulnificus*; *Aeromonas hydrophila*; *Plesiomonas shigelloides*; *Erysipelothrix interrogans*; *Pseudomonas* spp.; *Mycobacterium marinum*; *Leptospira interrogans*; *Clostridium botulinum*; *Giardia*; and *Diphyllobothrium*
Faecal pollution or contamination during processing and preparation	*Salmonella* spp.; *Shigella* spp.; *Escherichia coli*; *Campylobacter* spp.; *Listeria monocytogenes*; *Vibrio cholerae* O1; *Clostridium botulinum*; *Cl. perfringens*; *Bacillus cereus*; hepatitis A; unspecified hepatitis; Norwalk and Norwalk-like viruses

[a]Modified from Croonenberghs (2000), Dalsgaard (1998) and Ahmed (1992).

(Health Canada 1997a; CDC 1998a; CDC 1999a). Since these bacteria occur naturally in marine and estuarine waters, they are common in shellfish. For example, *Vibrio parahaemolyticus* levels in shellfish were found to be 200-fold higher than in overlying waters (DePaola et al. 2000). Their numbers are highly dependent on temperature in that water temperature is positively correlated with *V. parahaemolyticus* abundance and thus higher concentrations occur in summer months. However, no clear relation is found with salinity or fecal coliform levels. Like vibrios, isolation of *Plesiomonas shigelloides* is also a seasonal phenomenon. *P. shigelloides* is another opportunistic pathogen that can cause diarrheal disease and has been associated with the consumption of raw mollusks (Fernandes et al. 1997).

To reduce the risk for *V. parahaemolyticus* and other shellfish-associated infections, consumers should avoid eating raw or undercooked shellfish, especially during the warmer months. Monitoring of environmental conditions, such as water temperature, may help determine when shellfish harvesting areas should be closed to harvesting. During outbreaks, it is common for provincial or state health officials to close affected areas to shellfish harvesting. This has proven to be useful because subsequent infection rates decline. Guidelines regulating the harvesting of oysters and other shellfish rely on quantitative measurement of *V. parahaemolyticus* levels in shellfish meat. In Canada and the U.S., the recommended action level of *V. parahaemolyticus* per gram of shellfish meat that must be detected before closing shellfish beds is greater than 10,000 CFU/g. However, adherence to these guidelines does not prevent outbreaks. Recent outbreak investigations found that implicated oysters contained less than 200 CFU/g, indicating that human illness can occur at levels much lower than the current action level (Health Canada 1997a; CDC 1999a). More research is needed to confirm this observation.

As in all food, many pathogenic microorganisms are transmitted to humans via the fecal-oral route. Seafood may also be contaminated if it is harvested from water polluted with sewage. Shellfish in particular have been identified as vehicles for several viral gastroenteritis agents, such as Hepatitis A, Norwalk agent, and calicivirus (i.e. Norwalk-like or small round-structured virus). Outbreak investigations have often found inadequate sewage collection and disposal systems. Frequently, human waste from ill harvesters was the cause of contamination (CDC 1997b). The findings of these investigations suggested that one ill harvester could easily contaminate large quantities of oysters in a relatively large oyster bed. Oyster-associated outbreaks of viral gastroenteritis will continue unless regulators and the oyster industry develop, adopt, and enforce standards for the proper disposal of human sewage from oyster harvesting vessels. Traceback investigations also revealed the prevalence of mislabeling in wholesalers' records and thus there was an inability to accurately trace contaminated oysters to responsible harvesters. Therefore, prevention of oyster-related outbreaks of gastroenteritis not only requires educating workers in the oyster industry about the consequences of improper sewage disposal, but also improving record keeping by oyster harvesters, wholesalers, and retailers to enhance the reliability of traceback investigations.

Faecal pollution may also occur in aquaculture ponds and frequently involves ignorance of the hazards associated with the use of untreated animal or human waste in ponds to increase production (Garrett et al. 1997). The abuse and misuse of raw chicken manure as pond fertilizer may result in the transmission of *Salmonella* to the cultured product. Vibrios are also found naturally present in brackish water or the shrimp culture environment (Bhaskar & Setty 1994) and are one of the most dominant bacteria in the microflora of pond-reared shrimps. Seasonal variation is also a phenomenon in pond aquaculture environments and a major factor contributing to the association of pathogens with aquaculture products. *Salmonella* is generally not recognized as part of the normal flora in aquaculture environments and its presence in seafood is therefore regarded as a sign of poor standards of process hygiene and sanitation.

Dairy Products

Because milk is rich in nutrients and near neutral in pH, it serves as an excellent growth medium for many pathogenic and opportunistic microorganisms including enteric pathogens. Outbreaks and other illnesses involving milk and milk products have been documented since the beginning of the dairy industry (Table 1.3).

Table 1.3. Historical changes of pathogens in milk.[a]

Years	Diseases or pathogens
1900–1940s	Diphtheria, tuberculosis, brucellosis, Q-fever, septic sore throat
1950–1975	Salmonellosis, brucellosis, *Staphylococcus aureus*, *Bacillus cereus*, enteropathogenic *Escherichia coli*
1975–present	Enteropathogenic and enterohemorrhagic *E. coli*, *Yersinia enterocolitica*, *Campylobacter jejuni*, *Salmonella* Typhimurium, *S.* Derby, *S.* Dublin, *Listeria monocytogenes*, *Cryptosporidium parvum*

[a]Modified from Vasavada (1988).

Traditionally, raw milk and raw milk products are the major vehicles for transmission of enteric pathogens such as *Salmonella* Typhimurium, *S.* Derby, *S.* Dublin, and *Campylobacter jejuni* (Vasavada 1988). Raw milk is also a source for *Escherichia coli* O157:H7 and *Cryptosporidium parvum* (Karmali et al. 1983; Laberge et al. 1996). The incidence of outbreaks associated with dairy products has steadily declined due mainly to modern milk production practices that emphasize sanitary measures and the almost universal use of pasteurization. Commercial pasteurization was first introduced in 1895 and today milk is pasteurized both to destroy pathogenic bacteria that may be present and improve shelf life. In all territories and provinces including B. C., pasteurization of milk offered for sale is required by law (Health Canada 2002a). However, there are still uninformed individuals who maintain the belief that raw dairy products are healthier and pasteurized products are less beneficial and even harmful. Unfortunately, at the beginning of the 21st century, outbreaks of gastroenteritis due to consumption of raw milk and raw milk products continue to happen.

Raw milk can contain a diverse microbial population, which mainly comes from environmental contamination on farms, either during the milking process and subsequent storage, or from direct shedding by infected animals. Cows are carriers of some human enteric pathogens including *Listeria monocytogenes* and *Campylobacter jejuni*, and cows with mastitis can shed these pathogens in their milk (Vasavada 1988). Persistent shedding of *S.* Enteritidis from a cow into the milk has also been reported (Wood et al. 1991). Raw milk can also be contaminated with *E. coli* O157:H7 and *Salmonella* spp. from the environment. In a recent outbreak in B. C., unpasteurized goat milk was implicated as a vehicle for *E. coli* O157:H7 transmission in August 2001 (Health Canada 2002a). Pathogens present in raw milk can result in contaminated dairy products made from the contaminated milk. It is known that some enteric pathogens, including *Salmonella*, can survive the cheese-ripening process (El-Gazzar & Marth 1992). Therefore, consumption of raw milk and dairy products manufactured from raw milk carry potential risk. Cheese is a relatively uncommon vehicle for human *Salmonella* infection, but several outbreaks have been reported in recent years. Investigations often revealed raw milk used in the production of cheese as the source of contamination. In one outbreak of *Salmonella* Berta in Ontario in 1994, the implicated food was an unpasteurized soft cheese product produced on a farm and sold at a farmers' market (Ellis et al. 1998). Results of the investigation found that cheese was contaminated during ripening in an improperly disinfected bucket previously used for soaking chicken carcasses. This outbreak illustrated the potential role of uninspected home-based food producers in causing foodborne illness. *Listeria monocytogenes* has caused a number of dairy-associated outbreaks (Table 1.4). *L. monocytogenes* can survive the manufacturing process used for soft cheese manufacture (Health Canada 2002b). In February 2002, Abbott's Choice brand goat's milk cheese products from B. C. were recalled because some contained *L. monocytogenes* (Health Canada 2002c). Six illnesses were reported prior to the recall.

Dairy products made using pasteurized milk are essentially pathogen free and are seldom involved in outbreaks of foodborne illness. Over the past two

Table 1.4. Outbreaks associated with processed milk and milk products.[a]

Product	Pathogen(s)
Chocolate milk	*Yersinia enterocolitica, Staphylococcus aureus*
Pasteurized milk	*Yersinia enterocolitica, Listeria monocytogenes*[b]
Low fat milk	*Salmonella* Typhimurium
Dried milk products	*Salmonella* spp.
Cheese	*Staphylococcus aureus*
Mexican-style cheese	*L. monocytogenes*
Linderkranz cheese	*L. monocytogenes*
French Brie cheese	*L. monocytogenes*
Soft and semi soft cheese	*L. monocytogenes*, enteropathogenic *E. coli*, *S.* Berta (Ellis et al. 1998)
Cheddar cheese	*S.* Typhimurium (D'Aoust et al. 1985)
Mozzarella cheese	*S.* Oranienburg and *S.* Javiana (Hedberg et al. 1992)
Pasteurized cheddar cheese	*S.* Enteritidis (Ratnam et al. 1999)
Feta cheese (Abbott's)	*L. monocytogenes* (Health Canada 2002c)
Ice cream products	*L. monocytogenes*, *S.* Enteritidis (CDC 1994a)
Yogurt	*Escherichia coli* O157:H7 (Morgan et al. 1993)

[a]Modified from Vasavada (1988).
[b]Post-pasteurization contamination.

decades, however, pasteurized fluid milk and milk products have been implicated in several outbreaks of gastrointestinal illness (Table 1.3). In these incidents, improper pasteurization and post-pasteurization contamination were often the cause of the problem. In a major Canadian outbreak of *Salmonella* Typhimurium in 1984, deficiencies in the pasteurization procedure led to the addition of raw milk to pasteurized milk, which became the source of pathogens in the implicated cheddar cheese (D'Aoust et al. 1985). Over 2,700 people were infected in the outbreak. More recently, a nationwide outbreak of *Salmonella* Enteritidis phage type 8 associated with contaminated cheddar cheese occurred during March and April 1998 (Ratnam et al. 1999). The implicated cheese was contained in Schneider's Lunchmate prepackaged products. The outbreak resulted in nearly 700 reported cases, the majority of whom were children. It was one of the largest *Salmonella* outbreaks ever reported in Canada and the first time that pasteurized cheddar cheese was linked to a *Salmonella* Enteritidis outbreak. However the origin of the pathogen was not identified. Subsequent research has found that the *Salmonella* Enteritidis strain from the outbreak can survive the 60 day holding period required by Health Canada in the manufacture of cheeses from raw and pasteurized milk. Survivors were also detected in raw and pasteurized milk cheeses after 99 days of storage at 8°C (Modi et al. 2001). Even the most wholesome raw materials can be contaminated during processing. Poor sanitation in a cheese processing plant was the reported cause of a multistate outbreak of *S.* Oranienburg and *S.* Javiana in the U.S. in 1989 (Hedberg et al. 1992).

Many cheeses today are still made from unpasteurized milk (e.g. camembert cheese, aged cheddar, etc). The high fat content of cheese helps protect

Table 1.5. Fruit and vegetables implicated in recent outbreaks of enteric disease.[a]

	Date	Vehicle	Etiology	Reported cases	States/Provinces[b]
1	1990	Tomatoes	*Salmonella* Javiana	174	4
2	Apr.–Sept. 1990	Strawberries	Hepatitis A	51	2:GA, MT
3	July 1990	Salad (vegetable)	Hepatitis A	3	1:CA
4	Nov. 1990	Raw vegetables	*Giardia lamblia*	27	Not available
5	1991	Watermelon	*Salmonella* Javiana	39	Not available
6	June–July 1991	Cantaloupe	*Salmonella* Poona	>400	23/Can.
7	Aug. 1991	Coconut milk	*Vibrio cholerae*	4	1:MD
8	Oct.–Nov. 1991	Apple cider	*E. coli* O157:H7	23	1:MA
9	July–Aug. 1992	Salad (vegetable)	*Shigella flexneri*	46	2:MI, OH
10	Sept. 1992	Vegetables	*E. coli* O157:H7	4	1:ME
11	1993	Apple cider	*Cryptosporidium*	213	1:ME
12	1993	Tomatoes	*Salmonella* Montevideo	100	4
13	Mar. 1993	Garden salad	Enterotoxigenic *E. coli*	47	1:RI
14	July 1993	Pea salad	*E. coli* O157:H7	16	1:WA
15	Aug. 1993	Cantaloupe	*E. coli* O157:H7	27	1:OR
16	Aug. 1993	Salad bar	*E. coli* O157:H7	53	1:WA
17	Apr. 1994	Potatoes	*Clostridium botulinum*	30	1:TX
18	June–Aug. 1994	Scallions	*Shigella flexneri*	53	7
19	Fall 1994	Potato salad (unconfirmed)	*Listeria monocytogenes*	Not available	12
20	Mar.–July 1995	Alfalfa sprouts	*Salmonella* Stanley	242	17/Finland
21	May–Aug. 1995	Orange juice	*Salmonella* Hartford; *Salmonella* Gaminara; *Salmonella* Rubislaw	63	21
22	Fall 1995–Spring 1996	Alfalfa sprouts	*Salmonella* Newport	133 (OR; BC)	7/Can./Denmark
23	July 1995	Lettuce (leafy green, red, romaine)	*E. coli* O157:H7	>70	1:MT
24	Sept. 1995	Lettuce (romaine)	*E. coli* O157:H7	20	1:ID
25	Sept. 1995	Lettuce (iceberg)	*E. coli* O157:H7	30	1:ME
26	Oct. 1995	Lettuce (iceberg; unconfirmed)	*E. coli* O157:H7	11	1:OH
27	May–June 1996	Lettuce (mesclun; red leaf)	*E. coli* O157:H7	61	3:CT, IL, NY
28	May–June 1996	Raspberries	*Cyclospora cayetanensis*	1465	21/2:ON, QC
29	May–June 1996	Alfalfa and clover sprouts	*Salmonella* Montevideo; *Salmonella* Meleagridis	>600	1:CA

30	Aug. 1996	Lettuce	*Campylobacter jejuni*	14	1:OK
31	Sept.–Oct. 1996	Apple cider	*Cryptosporidium parvum*	31	1:NY
32	Oct. 1996	Apple cider	*E. coli* O157:H7	14	1:CT
33	Oct. 1996	Apple cider	*E. coli* O157:H7	6	1:WA
34	Oct. 1996	Apple juice	*E. coli* O157:H7	70	3:WA, CA,CO/1:BC
35	Feb.–June 1997	Sprouts (alfalfa and other varieties)	*Salmonella* Infantis; *Salmonella* Anatum	109	2:KS, MO
36	Mar.–Apr. 1997	Strawberries	Hepatitis A	236	1:MI
37	Mar.–Apr. 1997	Lettuce (mesclun)	*Cyclospora*	29	1:FL
38	Mar.–June 1997	Raspberries	*Cyclospora*	762	14/1:ON
39	June–July 1997	Basil	*Cyclospora*	305	3:MD, DC, VA
40	June–July 1997	Alfalfa sprouts	*E. coli* O157:H7	108	2:VA, MI
41	Aug. 1997	Cole slaw	Hepatitis A	44	1:MI
42	Sept. 1997–July 1998	Alfalfa and clover sprouts	*Salmonella* Senftenberg	52	2:CA, NV
43	Dec. 1997	Scallions	*Cryptosporidium*	55	1:WA
44	Dec. 1997	Lettuce (mesclun)	*Cyclospora*	12	5
45	Apr.–May 1998	Alfalfa sprouts	*Salmonella* Havana	18	2:CA, AZ
46	May 1998	Salad	*E. coli* O157:H7	2	1:CA
47	May 1998	Cole slaw	*E. coli* O157:H7	27	1:IN
48	June 1998	Alfalfa and clover sprouts	*E. coli* O157:non-motile	8	2:CA, NV
49	June 1998	Potato salad	*E. coli* O6:H16	6500	1:IL
50	June 1998	Fruit salad	*E. coli* O157:H7	40	1:WI
51	July–Aug. 1998	Parsley	*Shigella sonnei*	225	3:CA, MA, MN/2:AB, ON
52	Aug. 1998	Lettuce	*Shigella sonnei*	160	1:MN
53	Nov. 1998–Mar. 1999	Alfalfa sprouts	*Salmonella* Mbandaka	62	4
54	Dec. 1998–Feb. 1999	Mamey	*Salmonella* Typhi	13	1:FL
55	Feb.–Mar. 1999	Lettuce (iceberg)	*E. coli* O157:H7	72	1:NE

[a]Reprinted from CSPI (1999) with permission.
[b]Number of states (U.S.) or their identity or Canada (Can.)/other country.

pathogens from human gastric acidity, which increases the risk of human infection from contaminated cheese products. Since cheese is a ready to eat product, even a small amount of contamination poses risk to human health. Cultured dairy products such as yogurt were not thought to be vehicles of foodborne illness because of the acidity developed during fermentation that inhibits growth of unwanted microorganisms (Martin & Marshall 1995). Outbreaks of *E. coli* O157:H7 due to consumption of yogurt have proven this to be incorrect (Morgan et al. 1993). As such cultured dairy products should not be excluded from outbreak investigations.

Fruits and Vegetables

In the past 25 years, the consumption of fruits and vegetables has increased in Canada. Concomitantly, public health officials have reported an increase in the number of reported fruit and vegetable-associated disease outbreaks. Data collected between 1990 and 1998 in the U.S. showed that when foodborne illnesses are examined on the basis of the number of outbreaks, contaminated meats account for about 29%, produce (fruit and vegetables including juices and salads) account for about 24%, and seafood about 14% (Griffiths 2000). However, when the same data are analyzed based on the number of individual cases of illness, meat accounts for about 20%, produce for about 41% and seafood about 8%. This data demonstrated that there are a greater number of outbreaks involving produce than those involving meats. Recent outbreaks of enteric diseases associated with produce are presented in Table 1.5. Because they are raw agricultural commodities, fruit and vegetables are expected to harbour microorganisms, which occasionally are pathogenic. The lack of a lethal treatment during production, processing, and preparation increases the difficulty of eliminating the risk associated with consumption of these products. However, by identifying and controlling deficiencies, it is possible to reduce the risk of enteric illness among consumers of these products.

Fresh produce (fresh fruits and vegetables) can become microbiologically contaminated at any point from farm to fork. Preharvest contamination can occur in several ways including application of untreated or improperly composted animal manure as fertilizer, utilization of contaminated water for irrigation or mixing of pesticides, or fecal contamination from wild or domestic animals or from workers (FDA 1998). The presence of soil-borne pathogens on fresh produce is frequent and includes spores of *Clostridium* species (*Cl. botulinum* and *Cl. perfringens*), as well as spores of enterotoxigenic *Bacillus cereus*, and cells of *Listeria monocytogenes* (Beuchat & Ryu 1997). The presence of other pathogenic bacteria, viruses, and parasites in soil results largely from application of manure or untreated sewage, either by chance or design. Increased use of manure rather than chemical fertilizers may have played a role in increased numbers of produce-associated outbreaks. Untreated or improperly treated manure could introduce foodborne pathogens such as *Listeria monocytogenes, Salmonella* spp. and *Escherichia coli* O157:H7. Animal manure has been identified or suspected as a source of contamination of fresh produce or minimally processed produce that was associated with human disease outbreaks. Probably the most severe outbreak in Canada in terms of the consequences was the outbreak of *Listeria monocytogenes* that occurred in the Maritime Provinces in 1981 which resulted in 17 deaths (Schlech et al. 1983). Coleslaw was identified as the vehicle for disease transmission and was commercially prepared from contaminated cabbage. Investigation revealed that cabbage was grown on fields fertilized with composted and fresh sheep manure. Two sheep on the farm had previously died of listeriosis. After harvest, cabbage was kept in cold storage for four months before shipment to a coleslaw processor. Since *L. monocytogenes* is psychotrophic, the storage conditions had permitted the pathogen to multiply to an infectious level.

Newly emerged *E. coli* O157:H7 has been of particular concern due to its near ubiquitous presence in farm animals and environments. Cieslak et al. (1993) reported a community outbreak of *E. coli* O157:H7 due to consumption of vegetables from a garden where soil was regularly fertilized with cow and calf manure. An outbreak of *E. coli* O157:H7 infection in the U. K. was associated with simply handling, not consumption, of unwashed potatoes (Morgan et al. 1988). Several outbreaks of *E. coli* O157:H7 have been linked to consumption of unpasteurized apple cider. Probably the first recognized outbreak of *E. coli* O157:H7 in Canada occurred in 1980, where 13 Canadian children were diagnosed with hemolytic uremic syndrome (HUS) due to consumption of apple cider (Steele et al. 1982). Besser et al. (1993) attributed an outbreak of *E. coli* O157:H7 caused by consumption of fresh pressed apple cider to the inadequacy of washing and brushing of fruit prior to processing. The majority of apples were collected from windfalls, where apples may have had contact with cattle manure on the ground. These outbreaks imply that some traditional composting practices are insufficient to render animal manure safe for use on fruit and vegetables. Research has shown that *E. coli* O157:H7 can remain viable in bovine feces for up to 70 days (Wang et al. 1996), which means that regulations requiring aging of bovine manure for 60 days before using it as fertilizer may be inadequate (Tauxe et al. 1997). A recent research report demonstrated that *E. coli* O157:H7 from manure-contaminated soil and irrigation water could enter the lettuce plant through the root system and migrate throughout the edible portion of the plant (Solomon et al. 2002). These researchers also found that the pathogens migrated to internal locations in lettuce plant tissues and were therefore protected from the action of sanitizing agents simply by virtue of its inaccessibility. These findings agreed with those of Wachtel et al. (2002) who observed *E. coli* O157:H7 from contaminated irrigation water adhered preferentially to lettuce roots in both hydroponic and soil model systems, and to the deep grooves of seed coats in the hydroponic system. Significant bacterial numbers were associated with the edible portion of the lettuce plant. Their results strongly suggested that *E. coli* O157:H7 attachment to roots can cause contamination of edible root crops such as carrots, onions, and radishes, which have all been implicated in outbreaks of the pathogen.

E. *coli* O157:H7 outbreaks associated with apple cider have additional significance. Traditionally, foods with pH below 4.6 are not regarded as potentially hazardous. Cider typically has a pH of 3.5–4.0 due to the presence of malic and lactic acid. Hence the outbreaks have demonstrated the extraordinary acid tolerance of *E. coli* O157:H7. An outbreak of *Cryptosporidium* infection in 1993 was also linked to consumption of unpasteurized apple juice where dropped apples were collected from trees adjacent to an area grazed by cattle whose stool contained *Cryptosporidium* (Millard et al. 1994). These disease outbreaks have emphasized the need for heat treatment of apple cider, given the consequences of contaminated manure application.

Alternately, domestic sewage sludge can be used as an agricultural fertilizer. *Listeria* species have been reported in sewage, in particular *L. monocytogenes*. The organism was found to survive in sewage sludge for long periods of time; at least 13 months with minimal death during storage. Studies revealed treatment of sewage does not necessarily yield a pathogen free sewage sludge cake (Al-Ghazali & Al-Azawi 1988). *L. monocytogenes* survived activation and digestion stages, and was present at a low number, 3–15 cells per gram, in this material. These researchers later showed that soils treated with contaminated sewage sludge and crops grown on these soils became contaminated with the pathogen (Al-Ghazali & Al-Azawi 1990).

Irrigation and surface run-off waters can also be sources of pathogenic microbes that contaminate produce fields. Irrigation water containing raw sewage or improperly treated effluents from sewage treatment plants may contain hepatitis A, Norwalk and Norwalk-like viruses. The proximity of domestic or wild animals to surface water for irrigation can serve as a vehicle for *E. coli* O157:H7 to gain access to produce growing in the field. The organism has been shown to survive in pure water and lake water (see waterborne transmission). In addition, diluting some commercial pesticides with contaminated water has the potential to promote survival and growth of several human pathogens including *E. coli* O157:H7, *Salmonella* Typhimurium, *S.* Enteritidis, *Shigella sonnei*, *S. flexneri*, and *Listeria monocytogenes* and subsequently increase the risk of widespread contamination of produce (Guan et al. 2001). Wild birds are known to harbour some human pathogens, including *Campylobacter*, *Salmonella*, *Vibrio cholerae*, *Listeria* species, and more recently *E. coli* O157:H7. Pathogenic bacteria are likely to be picked up as a result of birds feeding on garbage, sewage, fish, or lands that are grazed with cattle or have had applications of fresh manure. Control of preharvest contamination of fruit and vegetables with pathogenic bacteria from wild birds would be extremely difficult. Contact of fruit and vegetables by infected field pickers and handlers at the time of harvest also offers a route by which pathogens can contaminate raw produce.

Postharvest contamination can occur as a result of using contaminated wash water or ice, improper handling, the presence of animals or other extraneous sources of pathogens in the processing environment, the use of contaminated equipment or transport trucks, and cross-contamination from other produce. In addition, processing practices that pool many individual fruits or vegetables increase the risk that a single contaminated item will contaminate the entire lot (Tauxe et al. 1997). During final food handling, basic handling hygiene and control of storage temperature, transport and display are all important.

Foodborne diseases associated with fresh produce often involve produce that has undergone some kind of minimal or non-thermal processing, such as fresh-cut fruit or fresh squeezed fruit juice, followed by time-temperature combinations that allow pathogens to survive and multiply to an infectious level. Normally, intact produce has an exterior physical barrier that prevents pathogens from entering into the nutrient-rich interior. The surface of produce is complex and can be difficult to sanitize, and pathogens can adhere tightly or hide in inaccessible locations. Once the surface integrity is broken, growth of pathogens can be rapid. Therefore, mechanical processing of fresh produce, such as cutting, shredding, or juicing, increases the risk of pathogen growth, and minimal heat treatment in fact makes produce more vulnerable to microbial penetration and multiplication. In view of these facts, the FDA Retail Foods Branch has designated fruit and vegetables that receive a heat treatment (scalding, blanching, or cooking) as potentially hazardous (Madden 1992).

WATERBORNE TRANSMISSION

Drinking water, water used in food production, for irrigation, for fish farming, waste water, surface water, and recreational water have all been recognized as vectors for transmission of human pathogens. Contaminated water has the potential to cause extensive outbreaks of illness due to the size of populations served by some distribution systems and the large number of people who use some recreational water facilities (Tillett et al. 1998). A waterborne disease outbreak is characterized by a gastroenteritis illness affecting a high proportion of a population, found in all age groups, and affecting persons with an epidemiological link to one community.

Drinking Water

Wholesome (clear, palatable and safe) drinking water is fundamental to public health. Drinking water including municipal or private water supplies and bottled water come from natural sources such as groundwater, wells, springs, and glaciers (Warburton & Dodds 1992). There are occasions when even water from natural sources may be contaminated. Many subsurface environments, including shallow and deep aquifers down to 4000 m, actually contain a wide variety of microorganisms, sometimes with a substantial population of as many as 10^5–10^7 CFU per ml (Warburton and Dodds 1992). U.S. regulatory agencies suggest that bacterial counts in finished drinking water should not exceed 500 CFU/ml in order to reduce the interference with the detection of coliform bacteria (Environmental Protection Agency 1987 cited by Payment 1989). Coliform and *E. coli* (or fecal coliform) are the most common microbiological tests done on water. They are used as an indicator of water pollution. *E. coli* (as an indicator of fecal pollution) must not be present in 100 ml samples of any water intended for drinking (WHO 1993). This criterion is readily achievable by water treatment. When coliform bacteria are detected in treated water supplies, it suggests inadequate treatment, post-treatment contamination, or excessive nutrients (WHO 1993). More recent work, however, suggests that gastroenteritis is more strongly associated with the presence of *Enterococci* than of *E. coli* (Barrell et al. 2000).

Microbial populations in drinking water can either be indigenous or contaminants. Indigenous microorganisms in drinking water include *Acinetobacter, Cytophaga, Flavobacterium, Moraxella, Pseudomonas,* and*Xanthomonas* (Warburton & Dodds 1992). These microorganisms are not formally recognized as pathogens, thus do not pose any public health risk when found in drinking water; although at high number they have the potential to cause disease especially in persons with impaired defense mechanisms. In addition to indigenous bacteria, natural source water may contain a wide range of saprophytic species as well as human pathogens. Drinking water is a known vehicle for enteric pathogens such as *Aeromonas, Pseudomonas, Yersinia, Campylobacter, Escherichia, Salmonella, Shigella, Vibrio, Giardia, Cryptosporidium, Cyclospora,* hepatitis A, Norwalk, and Norwalk-like viruses (Warburton & Dodds 1992; Todd et al. 2000). Waterborne disease outbreaks are a result of either poor protection of the source water or inadequate water treatment. Use of contaminated, untreated groundwater caused 35% of waterborne disease outbreaks during a 30-year period in the U.S. (Warburton 1993).

In Canada, municipalities with contaminated raw drinking water and minimal water treatment have been found in all regions (Wallis et al. 1996). The outbreak of *Escherichia coli* O157:H7 and *Campylobacter* infections in Walkerton, Ontario, in the summer of 2000 marked the first documented outbreak associated with a treated municipal water supply in Canada and the largest multi-bacterial waterborne outbreak in Canada to date (Health Canada 2000c). The outbreak resulted in 1,346 reported cases and 6 deaths (Health Canada 2000c; Jones 2000). Heavy rainfall was believed to be responsible for gross contamination of the distribution system. Investigations revealed heavy rain accompanied by flooding, pathogens present in cattle manure from adjacent farms, a well subject to surface water contamination, and a water treatment system that may have been overwhelmed by increased turbidity, contributed to the unfortunate consequences. The outbreak has prompted questions about the safety of groundwater sources that may be under the influence of surface water, especially under flood conditions. Historically, groundwater sources have been assumed to be secure and consequently are treated with chlorination only. However, in light of the Walkerton tragedy, the approach needs to be re-evaluated. Such evaluation should take into account all current and future pressures on land use including human population density and agricultural activities. Ogden et al. (2001), who studied transport of *E. coli* and *E. coli* O157:H7 after cattle slurry application on drained plots in grassland and arable stubble, reported leaching losses of 0.2 to 10% of total *E. coli* and leaching was dependent on rainfall. Risk of significant pollution of water by *E. coli* was highest immediately after slurry application, and the first increments of drain flow carried significant concentrations. Wang et al. (1996) reported that *E. coli* O157:H7 could survive for long periods of time, up to 91 days in water especially at low temperatures. Nonetheless, *E. coli* O157:H7 is highly susceptible to chlorine, with more than a 7 log reduction by 0.25 ppm free chlorine within 1 min (Zhao et al. 2001). Therefore, inadequate water treatment remains the major cause in waterborne outbreaks of *E. coli* O157:H7. Ever since its first isolation from water (McGowan et al. 1989) and first waterborne outbreak (Dev et al. 1991), waterborne disease from *E. coli* O157:H7 continues to recur globally (Swerdlow et al. 1992; Isaacson et al. 1993; Jackson et al. 1998; Licence et al. 2001; CDC 1999b), mainly due to the use of untreated, unprotected private water sources in areas where animal activities are high.

Contamination of well water has also been implicated in outbreaks of other bacterial pathogens such as *Shigella sonnei* (Lindell & Quinn 1973; CDC 1996c; Alamanos 2000), pathogenic *E. coli* (UGPR 1997), *Campylobacter* (CDC 1999b), *Plesiomonas shigelloides*, and *Salmonella* Hartford (CDC 1998b). A 1991/92 Ontario farm groundwater quality survey found that 20% of private wells exceeded the Ontario drinking water objectives for fecal coliform bacteria (UGPR 1997). Contamination of wells tends to be episodic in nature and is more prevalent in the summer than in the winter. Intermittent contamination problems are caused by the overflow or seepage of sewage, surface water runoff, and periodic flooding over the well field and can be exacerbated by poorly maintained wells. In addition, lateral migration of microorganisms can occur over considerable distances (2 to 3 km) especially in unconfined sandy aquifers and is more substantial in ground water (Warburton 1993).

In recent years, waterborne parasites such as *Giardia*, *Cryptosporidium*, and *Cyclospora*, have risen to be the major diarrheal problems in drinking water in North America, with *Cryptosporidium* and *Giardia* being much more frequent than *Cyclospora*. In fact, giardiasis is the most common parasitic infection. Giardiasis and cyclosporiasis are treatable but cryptosporidiosis is not (Rose & Slifko 1999). Giardiasis can last from a few days to a week or, on rare occasion, months. Cryptosporidiosis is cholera-like, usually accompanied by large volumes of fluid loss, and can last from 3 to 12 days in immuno-compromised persons, and occasionally more than 2 weeks (Current & Garcia 1991). Cyclosporiasis is less severe in the immunocompetent and usually lasts from 3 to 25 days, but can last for several months in an immunocompromised host. *Cyclospora* oocysts require a maturation period, but *Cryptosporidium* oocysts and *Giardia* cysts are immediately infectious upon excretion from host. Both *Cryptosporidium* and *Giardia* are associated with cross transmission from animals to humans whereas *Cyclospora cayetanensis* has only been documented in humans and baboons (Rose & Slifko 1999). Calves, companion animals such as rodents, puppies, and kittens, and their wastes are a source of *Cryptosporidium* (Current & Garcia 1991) while beavers have been implicated in waterborne outbreaks of *Giardia* (Rose & Slifko 1999). *Cryptosporidium* oocysts and *Giardia* cysts can persist and survive for long periods of time in water. Research has shown that *Cryptosporidium* oocysts survived up to 176 days in drinking or river water with inactivation of 89–99% (Robertson et al. 1992), while *Giardia* cysts remained viable for up to 56 days in river water with 75–99.9% inactivation (DeRegnier et al. 1989). *Cryptosporidium* oocysts also survive well in human and cattle feces as fecal material protects them from desiccation, thus prolonging their viability within the environment. *Cyclospora* infection may have seasonality. Studies found that the greatest maturation of oocysts occurred at a temperature of 30°C when compared with maturation at 4 and 37°C (Smith et al. 1997). These protozoa are, however, sensitive to extreme temperatures, such as heating and freezing, which destroy the viability of oocysts and cysts (Rose & Slifko 1999).

Cryptosporidium parvum oocysts are very resistant to most disinfectants including chlorine (Current & Garcia 1991). Consumption of untreated surface water is a predominant risk factor for acquiring cryptosporidiosis. The first documented waterborne outbreak of this organism in North America occurred in San Antonio, Texas in 1986, and was linked to sewage leakage into well water. The well water was chlorinated but not filtered. Wastewater in the form of raw sewage and runoff from dairies and grazing lands were identified as sources of oocysts that contaminated drinking water. The importance of agricultural sources of oocyst contamination should not be taken lightly since infected calves and lambs can pass up to 10^{10} oocysts per day for up to 14 days. Therefore, large numbers of oocysts can enter the surface water system following a hard rain on a pasture containing infected animals. A recent waterborne outbreak of cryptosporidiosis occurred in North Battleford, SK, in the spring of 2001 and affected approximately 7,000 people (Health Canada 2001b). *Cryptosporidium parvum* was found in finished municipal drinking water. The cause of the problem was a deficiency of an operational parameter in the treatment plant that received water from a surface river. The source of oocysts was not found, but was assumed to come from some point upstream of the water intake from the river. Probably the largest documented outbreak of waterborne disease in North America occurred in Milwaukee, Wisconsin, in the spring of 1993 where 400,000 people became ill and approximately 100 people died (Mac Kenzie et al. 1994). The massive outbreak was attributed to *Cryptosporidium*. Except for the increased turbidity in the treated water, water quality measurements were within specified limits and there was no evident mechanical breakdown of flocculators or filters in the treatment plant. It seemed that *Cryptosporidium* oocysts passed through the filtration system of the water treatment plant. The outbreak demonstrated that water quality standards followed and testing used failed to eliminate the organism. It was speculated that possible sources included cattle along the rivers that flow into Milwaukee harbour, slaughterhouses, and human wastes. Rivers that were swelled by spring rains and snow runoff may have transported oocysts into Lake Michigan and from there to the intake of the treatment plant. *Cryptosporidium* is associated with diarrheal illness in most areas of the world including developed countries, indicating the continuing risk of cryptosporidiosis from chlorinated water supplies (Furtado et al. 1998). An interesting finding by McAnulty et al. (2000), who reported an outbreak of *Cryptosporidium* caused by contaminated municipal water in the absence of a discernible outbreak among the residents, suggested that persons previously exposed to contaminated water may develop immunity to cryptosporidiosis. Prevalence of cryptosporidiosis is usually higher in children (especially those less than 2 years of age), than in adults, and small numbers of oocysts may be present in feces for up to 2 weeks following resolution of diarrhea (Current & Garcia 1991). Oocysts have been found in 5.6–87.1% of source waters sampled in the U.S. and Canada, and in 3.8–40.1% of drinking water samples in the U.S., Canada, and Scotland (Rose et al. 1997 cited by Davis et al. 1998). In environments across Canada, Wallis et al. (1996) found oocysts in 6.1% of raw sewage, 4.5% of raw water, and 3.5% of treated water samples. These surveys have confirmed the ubiquity of *Cryptosporidium* oocysts.

Giardia is a common enteric parasite in Canada. There have been several documented outbreaks of giardiasis in British Columbia, Alberta, Ontario, Quebec, New Brunswick, and Newfoundland (Wallis et al. 1996). *Giardia* cysts are commonly found in raw surface waters and sewage in Canada. In a Canada wide study, *Giardia* cysts were found in 73% of raw sewage samples, 21% of raw water samples, and 18.2% of treated water samples (Wallis et al. 1996). There was a trend to higher concentration and more frequent incidence of cysts in the spring and fall, but positive samples were found in all seasons. High concentrations of cysts in raw sewage and positive results from drinking water should be a concern to public health authorities. The ubiquity of cysts in sewage indicates that when drinking water is positive, untreated sewage has been discharged into surface water or the water source is inhabited by aquatic mammals such as beaver and muskrat that may have become infected with *Giardia* from human sources. Establishment of *Giardia* spp. in wild animal reservoirs helps to spread the parasite through animal migration and may magnify an initial low level of contamination to potentially dangerous concentrations. The inoculum required to infect humans was found to lie between 10 and 100 cysts (Rendtorff 1978 cited by Wallis et al. 1996).

In summary, transmission of waterborne parasites through the environment is facilitated by several factors: (1) oocysts and cysts are stable for weeks to months in the environment; (2) low infectious dose—one oocyst or cyst of *Cryptosporidium* and *Giardia*, respectively, can initiate infection in humans; (3) most feces carrying oocysts and cysts end up in the environment and can be spread to foods such as produce by irrigation or by direct contact; and (4), routine wastewater treatment eliminates only a small fraction of the oocysts and cysts (Rose & Slifko 1999). Proper filtration should remove oocysts and cysts; *Giardia* cyst is 12 by 6 µm, whereas *Cryptosporidium* and *Cyclospora* oocysts are 4–5 µm and 8–10 µm, respectively.

It is not unusual that the causative agent is not isolated from the implicated water in an outbreak. Detection and recovery of indicator microorganisms as well as pathogens from environmental samples are often technically difficult because of the altered physiological state the microorganism must exist in to survive the hostile environment, and the low numbers in which they occur (Payment et al. 1989). In addition, laboratory tests usually require samples of large volume and techniques and methods used may not be appropriate and sensitive enough. Further, contamination of the water supply is likely to be intermittent.

Watershed protection and chlorination together do not necessarily eliminate all risk of waterborne disease transmission (Health Canada 2000d). Oocysts and cysts are resistant to most types of chemical disinfection. Drinking water treatment can considerably reduce the level of viruses but may not eliminate them completely. Several studies have shown that coliform-free drinking water may still contain viruses or parasites. In Canada, changes in water quality guidelines and regulations are being driven by the example set by the U.S. Environmental Protection Agency (1989, 1994 cited by Payment et al. 2000). The maximum contamination level for viruses and parasites has been set to zero and this should be achieved through proper treatment. Regulations require all systems using surface or ground water under the influence of surface water to disinfect the distributed water. Systems using surface water must treat the water to remove or inactivate at least 99.9% (3 log reduction) of *Giardia* cysts, and at least 99.99% (4 log reduction) of enteric viruses. A higher level of treatment could be required for waters that are more contaminated. Minimal treatment for source water should include filtration for cyst removal, in series with disinfection for pathogenic bacteria and viruses. However, it is of interest that until recently most of B. C. drinking water supplies were still obtained from unfiltered surface sources (Isaac-Renton et al. 1996). It should also be noted that water used for making edible ice should be subject to the same drinking water standard and include specific sanitary requirements for equipment for making and storing ice (WHO 2000). The same requirements should also be extended to ice used for storing and transporting food.

The mode of transmission of pathogens through water includes direct exposure via ingestion of, or contact with contaminated water, and indirect exposure through ingestion of foods or products prepared with contaminated water, subsequent contact with infected persons or animals, and exposure to aerosols (Warburton & Dodds 1992). Contaminated water supplies might be a more significant source of infection than foods, especially for infants (Warburton 1993). Intestinal problems caused by contaminated water supplies often go unreported, since the problems may be short-lived and are probably not associated with the direct ingestion of water. If food is contaminated by water containing pathogens that multiply in the food, or if a susceptible person becomes infected by water, subsequently infecting others by person-to-person contact, the initial involvement of water may be unsuspected (WHO 1993). Secondary contamination from individuals infected through the water route also constitutes a potential source of infection and contributes to the endemic level of illness in the population (Payment et al. 2000).

Recreational Water

Enteric disease outbreaks associated with recreational water activities have been documented in the past 20 years (Table 1.6). Two bacterial pathogens, *Escherichia coli* O157:H7 and *Shigella*, and two pathogenic protozoans, *Giardia* and *Cryptosporidium*, are of particular interest because of the frequencies and circumstances under which the outbreaks occurred. These outbreaks usually occurred in small, shallow, stagnant bodies of water and involved mostly children.

Table 1.6. Swimming-associated disease outbreaks.

Year	Place	Causative agent	Reference
1984	Public pool, USA	*Giardia*	Harter et al. (1984)
1985	Public pool, USA	*Giardia*	Porter et al. (1988)
1985	Artificial lake, USA	*Shigella sonnei*	Sorvillo et al. (1988)
1988	Water slide pool, Canada	*Giardia*	Greensmith et al. (1988)
1988	School pool, USA	*Cryptosporidium*	CDC (1990)
1988	Pool at sport center, UK	*Cryptosporidium*	Joce et al. (1991)
1989	Pond, USA	*S. sonnei*	Blostein (1991)
1989	Public pool, USA	Hepatitis A	Mahoney et al. (1992)
1990	Public pool, Canada	*Cryptosporidium*	Bell et al. (1993)
1991	Lakeside park, USA	*Escherichia coli* O157:H7 & *S. sonnei*	Keene et al. (1994)
1992	Wave pool, USA	*Cryptosporidium*	McAnulty et al. (1994)
1992	Paddling pool, Scotland	*E. coli* O157:H7	Brewster et al. (1994)
1993	Public pool, USA	*Cryptosporidium*	CDC (1994b)
1993	Paddling pool, UK	*E. coli* O157:H7	Hildebrand et al. (1996)
1994	Freshwater lake, USA	*E. coli* O157:H7	Ackman et al. (1997)
1995	State park lake, USA	*E. coli* O157:H7	CDC (1996d)
1997	Freshwater lake, Finland	*E. coli* O157:H7	Paunio et al. (1999)
1997–1998	Public pool, Australia	*Cryptosporidium*	Puech et al. (2001)

Epidemiological investigations of the outbreaks revealed that the source of the causative agents was usually the bathers themselves.

S. sonnei and *E. coli* O157:H7 are two closely related organisms that have been linked to outbreaks of illness associated with swimming. In most outbreaks, they were the singular causative agent. But in one outbreak that occurred in Oregon in 1991 (Keene et al. 1994), both organisms were responsible, which underscores the fact that these organisms share many characteristics, epidemiological as well as biological and clinical. *S. sonnei*-associated outbreaks usually involved a natural lake or pond whereas outbreaks of *E. coli* O157:H7 have occurred in both lakes and swimming/paddling pools. In virtually all outbreaks, the infected swimmers fecally contaminated the water and thereafter spread the infection to other bathers. In some cases, toddlers and bathers that were not toilet trained and had diarrhea visited the swimming area. Fecal contamination is even more likely to occur when there is a high density of bathers. Swallowing this fecally contaminated water is the primary mode for transmission of enteric pathogens in recreational water outbreaks. Because young children are usually not skilled in swimming, they are more likely to swallow pool water while swimming. Infection after the incidental ingestion of small amounts of water is most likely to occur if the infectious dose is low, which is well recognized in shigellosis and *E. coli* O157:H7 infection. In addition, *S. sonnei* and *E. coli* O157:H7 are known to survive in lake water for weeks or months. A prolonged outbreak in Oregon (Keene et al. 1994) that continued for more than 3 weeks, confirmed both the lengthy survival of these enteric pathogens in lake water and the low infectious dose. In a small community outbreak (Brewster et al. 1994), a paddling pool that had not been drained and had no disinfectant added was implicated in *E. coli* O157:H7 transmission to children who visited the pool over a 3 day period. Re-use of contaminated paddling pools over a period of several days clearly poses a potential risk of transmission of enteric infection. Swimming-associated outbreaks all occurred during hot summer days when the water temperature, swimming activity, and bather density were high. The risk is intensified when the water bodies are small. In most outbreaks, causative agents were not isolated from the implicated water but the numbers of indicator organisms were high. In contrast, because the infectious dose of some enteric pathogens is very small, recreational water that meets current standards for indicator organisms may still have a sufficient number of pathogens to cause infection (Ackman et al. 1997). Further, indicator organisms have been shown to correlate poorly with rates of gastroenteritis. Current methods for collecting and testing recreational water may be insensitive to low concentrations of pathogens such as *E. coli* O157:H7. The intermittent nature of *E. coli* O157:H7 contamination limits the ability of microbiological and environmental monitoring to detect its presence. Because routine monitoring of recreational water may not detect pathogen contamination, primary prevention remains the ultimate method for elimination of fecal contamination from swimming areas. The solution is reliant more on public cooperation than government regulation. Children who are not toilet trained pose a danger to others at swimming areas.

Risk of enteric illness associated with swimming that has been linked to pathogenic protozoa mainly involves two parasites: *Giardia* and *Cryptosporidium*. These two organisms share a number of characteristics. They produce a cyst or oocyst that is environmentally stable and highly resistant to disinfectants. They have a low infective dose and they are shed at high densities by infected individuals. There have been a number of outbreaks attributed to these pathogens (Table 1.6). Outbreaks caused by these organisms are usually associated with swimming in chlorinated water, such as swimming pools, water slides, and wave pools, and involve both adults and children. The sources of parasite contamination in the pools were either the swimmers themselves or sewage.

Disinfection of pools seems to have little effect upon disease outbreaks caused by these parasites. In a case of giardiasis in 1984 (Harter et al. 1984), the patient participated in an infant and toddler swim class where adequate chlorine levels were maintained in the pool. Contamination was believed to be due to accidental fecal release by class members. In 1988, an outbreak of 60 cases of cryptosporidiosis was reported in Los Angeles County (CDC 1990), where the implicated swimming pool contained adequate chlorine levels (2 ppm), but had a 30% diminished filtration flow rate. In one outbreak of giardiasis that occurred in an indoor pool (Porter et al. 1988), the chlorine level was zero on the day after an incident of accidental fecal release by a swimmer. Seven out of nine patients were adults. In Manitoba, an outbreak of giardiasis was associated with a hotel's new water slide pool (Greensmith et al. 1988). Patients ranged from 3 to 58 years of age. The close proximity of a toddlers' wading pool was thought to be the source of fecal material in the water slide pool. Inadequate pool maintenance is also an important contributing factor. In August 1988, an outbreak of cryptosporidiosis occurred in a swimming pool at a local sports center in the U.K. (Joce et al. 1991). Oocysts were detected in the pool water, indicating heavy contamination. Outbreak investigations revealed significant plumbing defects, which had allowed inflow of sewage from the main sewer into the circulating pool water. Because oocysts and cysts can remain infectious for several months in moist environments, outbreaks can potentially be prolonged.

In conclusion, although chlorine is an effective disinfectant for bacterial pathogens, it does not kill all pathogens, particularly the parasites. During 1989–1998, approximately 10,000 cases of diarrheal illness were associated with 32 recreational waterborne disease outbreaks in disinfected water venues in the United States (CDC 2001b). *Cryptosporidium* is highly resistant to the chlorine concentration routinely used in pools. *Giardia* is only moderately chlorine sensitive. Due to their small size, oocysts and cysts may not be adequately removed by the usual pool filtration system (Puech et al. 2001). Because of frequent fecal contamination, either accidental or intentional, the inability of chlorine disinfection to rapidly inactivate several pathogens and the common occurrence of accidental ingestion of pool water especially among young children, transmission of pathogens can occur even in well-maintained pools. Therefore, the cooperation of pool users and pool operators remains paramount in control and prevention of these disease outbreaks.

AIRBORNE TRANSMISSION/OCCUPATIONAL EXPOSURE

Food Processing Plants, Sewage Treatment Plants, and Farm Environments

Air contains gases, water droplets, suspended particles of pollen and dust, and microorganisms (bacteria, viruses, yeasts, and molds) which may be regarded as aerosols (Al-Dagal & Fung 1990). An aerosol is defined as a solid or liquid particle suspended in air (Zottola et al. 1970). Particulates in an aerosol usually vary in size from <1 µm to approximately 50 µm or possibly larger. These particulates may consist of a single organism or a clump of a number of organisms. Microorganisms adhere to a dust particle or exist as a free particle surrounded by a film of dried organic or inorganic material (Zottola et al. 1970). Microorganisms cannot grow in the air due to the absence of nutrients, but they can be transmitted for long distances. Major sources of airborne microbes include humans (through sneezing, coughing, and talking), animals, sewage, and dust particles that act as absorbents of both microbes and endotoxins.

Fecal wastes and other discharges from humans are the most important sources of indirect air contamination. One astonishing example is a recurrent outbreak of viral gastroenteritis on a cruise ship (Ho et al. 1989). Apparently over a 2-year period, there were more than 10 outbreaks of diarrhea on this cruise ship despite repeated investigations and extensive efforts to control the epidemic with traditional approaches, namely, provision of clean water and food. Results of investigations indicated that person-to-person contact or aerosolized droplets were likely modes of transmission. Epidemiological findings implicated vomitus in the transmission of viral diarrhea. A 32-nm small round structured virus (SRSV), possibly related to the Snow Mountain agent, was implicated as the cause of the recurrent outbreaks. Therefore, droplet or airborne spread of the Norwalk-like agents or any disease agent of low infectious dose would be of particular importance because vomiting is a prominent symptom and occurs in 50–90% of patients with acute gastroenteritis. In most outbreaks of acute gastroenteritis, a causal agent cannot be identified. Airborne, droplet, or contact spread of Norwalk and Norwalk-like agents may have contributed to outbreaks in which there was no common exposure to contaminated food or water.

Animals contribute to aerobiological pollution directly by creating air turbulence while moving in the field or indirectly by their fecal materials and other discharges (Al-Dagal & Fung 1990). Birds can carry microbes on their feathers and release them into the atmosphere in a processing plant, creating occupational hazards for plant workers. Zottola et al. (1970) indicated that in turkey processing plants, *Salmonella* are frequently found in the air of areas where live birds are handled. Gast et al. (1999) found that *Salmonella* Enteritidis contamination in commercial broiler hatcheries could be reduced effectively by minimizing dust, fluff, and aerosol production. *S.* Enteritidis is capable of surviving in aerosols for long periods of time, implying that there are potential health hazards from human exposure to contaminated air (Seo et al. 2001). Contamination from airborne microbes varies among different food processing plants. For instance, the airborne microbial contamination in a pork-processing establishment was higher by a factor of 10 than in dairy plants (Kotula & Emswiler-Rose 1988). Flying birds and insects are responsible for transmission of microbes from place to place and for redistribution of the airborne microbial population (Al-Dagal & Fung 1990).

Sewage treatment plants that use activated sludge, trickling or rock filter systems, create aerosols and droplets containing microbes including pathogens that contaminate the environment. As many as 10^5 CFU/m^3 of bacteria were found at a commercial sludge application site and indicators such as H$_2$S producers and pathogenic clostridia were present in locations of significant physical agitation of the sludge material (Pillai et al. 1996). In an early study by Teltsch & Katzenelson (1978), aerosolized coliforms including *E. coli* were detected when the concentration was 10^3/ml or more in wastewater. They reported a positive correlation between relative humidity and the number of aerosolized bacteria, and a negative correlation between solar irradiation and bacterial level in the air. During night irrigation, 10 times more aerosolized bacteria were detected than with day irrigation. Echoviruses were isolated in air samples collected 40 m downwind from the sprinkler. An investigation of aerosols emitted by trickling filter sewage treatment plants (Adams & Spendlove 1970) also revealed that coliforms were indeed emitted and could be sampled at a distance of 1.2 km downwind. It is apparent that the aerosolized coliforms arising from sewage treatment plants may portend a public health concern especially for the employees. Heng et al. (1994) showed that sewage workers in Singapore have an increased occupational risk of acquiring hepatitis A virus infection. Seroprevalence was 2.2 times higher than that of another non-occupationally exposed population, and a recommendation was made that they should be protected by active immunization.

Agricultural work often involves life-long exposure to dusts at high concentrations (Thorne et al. 1992). These high dust exposures occur especially during harvesting, grain and silage handling, and work in livestock buildings. Airborne microbes are known to cause pulmonary diseases associated with inhalation of agricultural dusts. Marshall et al. (1988) studied the survival of *E. coli* after aerosol dispersal in a laboratory and a farm environment, and reported that in both environments, the number of airborne bacteria declined rapidly within the first 2 h. However longer survival was found in the barn than in the lab (20 d versus 1 d). The longer survival in the barn was believed to be due to the higher humidities (61 to 82%). These studies indicated that *E. coli* can persist in the barn environment for an appreciable period of time and thus pose a farm health concern. An outbreak of *Campylobacter jejuni* infection occurred among farm workers in Ontario in 1994 (Health Canada 1995). The casual farm workers (one-day job) were employed to catch and transport a total of 13,000 six-week-old turkey poults from a brooding barn to a growing barn. The cases reported eating and smoking as they worked. Hand washing facilities were not available, and the workers were not advised about the risks of pathogenic bacteria shed by the birds nor the need for thorough hand washing before eating or smoking. Although the exact mode of transmission cannot be confirmed, the outbreak underscores the occupational hazards posed by zoonotic disease agents that often go underestimated. This outbreak illustrates the need for a comprehensive educational package on zoonotic disease prevention on farms.

Studies by Nijsten et al. (1994) reported that *E. coli* isolated from fecal samples of hog farmers showed the highest percentages of resistance to 5 antibiotics (amoxicillin, neomycin, oxytetracycline, sulfamethoxazole and trimethoprim) as compared to abattoir workers and (sub)urban residents. This observation illustrates another uncharacterized hazard associated with animal rearing.

DIRECT CONTACT TRANSMISSION

Animal-to-Person Transmission

As mentioned, pathogens that are capable of being transmitted directly from animals to humans are named zoonoses. Outbreaks of enteric infection due to zoonoses have often implicated farm visits as a risk setting. Most victims of these outbreaks are children. In most if not all cases, the implicated farm did not have appropriate handwashing facilities and proper warning signs about the risk of touching animals.

In Canada, the first reported outbreak of *Escherichia coli* O157:H7 infection associated with a petting zoo (open farm) occurred in Middlesex-London,

Ontario, in September–October 1999 (Warshawsky 2001). No common food source was identified, but all seven patients reported having touched animals in the Agriculture Pavilion at the Western Fair. Outbreak investigations further revealed 62 people that met case definition, with a median age of 7 years. In a subsequent case-control study, cases were more likely than controls to have touched goats at the petting zoo (Henry et al. 2001). Lack of handwashing and length of time in the petting zoo were also associated with illness. Six cases and one goat from the petting zoo were laboratory confirmed with *E. coli* O157:H7 phage type 27, a very rare subtype in Canada, all with the same PFGE pattern. After this outbreak, the Ontario Health Unit imposed local control measures including mandatory on-site handwashing facilities at petting zoos.

In the United States, the first reported outbreaks associated with direct transmission of *E. coli* O157:H7 from farm animals to humans occurred among school children in Pennsylvania and Washington during the spring and fall of 2000 (CDC 2001c). The outbreaks were associated with school and family visits to farms where children came into direct contact with farm animals. The CDC survey in June 2000 revealed that of the 44 state and territorial public health departments, none had laws to control exposure of humans to enteric pathogens at venues where the public had access to farm animals, and no federal laws existed that addressed this public health issue. Following these farm visit-related outbreaks, CDC, the Zoonoses Working Group, the National Association of State Public Health Veterinarians, the U.S. Department of Agriculture, the Animal and Plant Health Inspection Service, and other groups, drafted measures to reduce the risk for farm animal to human transmission of enteric infection.

In the United Kingdom, farm-associated outbreaks of *E. coli* O157:H7 infections have been reported among children (Shukla et al. 1995; Milne et al. 1999; Chapman et al. 2000). In one outbreak, *E. coli* O157:H7 strains indistinguishable from those isolated from the human cases, were isolated from composted mix of animal manure and vegetable waste which had been processed for 3 months, but not from partially finished compost which had been processed for only 6 weeks (Chapman et al. 2000). There has been no report on the survival of the pathogen during the composting of mixed animal manure and vegetable waste. Chapman and coworkers suggested if *E. coli* O157:H7 survived the 3 months of composting, the pathogen could have multiplied to large numbers in the final composted mix. This could explain why the pathogen was only detected in finished compost (3 months) but not in partially finished compost (6 weeks) where the number of pathogens might be very low. Another outbreak of *E. coli* O157:H7 infection was associated with a music festival in England in 1997 (Crampin et al. 1999). Investigations identified no common food or water source but all cases reported a high level of mud contamination, especially on hands and faces. Subsequently, undistinguishable *E. coli* O157:H7 strains were isolated from all cases and from a cow belonging to a herd that had previously grazed the site of the festival, suggesting that mud contaminated with cattle feces was the likely vehicle of infection. As a consequence of this outbreak, the local authority has ensured better hygiene recommendations are emphasized in future festivals. They recommended that cattle be excluded from the site 2 weeks beforehand and that the site be chain harrowed to facilitate fecal decomposition. In a case control study of sporadic *E. coli* O157:H7 infection in England (O'Brien et al. 2001), exposure to the farming environment emerged strongly as a risk factor. The risk occurred in people not routinely exposed to the farming environment, i.e., public visiting open farms or spending holidays on farms, or people going to farms for work but not regularly employed on farms. With respect to recreational visits, half of the patients reported touching farm animals, the remainder had simply been exposed to the environment. These findings are consistent with previous descriptive studies undertaken in Scotland by Coia et al. (1998) who demonstrated that farmers who routinely worked with livestock were not found to be at increased risk. These same conclusions were reached in a Canadian study described below.

Because dairy farm families are directly exposed to verocytotoxigenic *E. coli* (VTEC) through personal contact with cattle manure and through consumption of raw milk, both of which are known risk factors for VTEC infection, they constitute a model of naturally occurring transmission of these pathogens from cattle to people. In a study by Wilson et al. (1998) of dairy farm families in southern Ontario, a lack of disease due to exposure to VT1-producing *E. coli* in farm residents suggested enhanced protection occurred due to antibodies induced by previous exposure. They emphasized that exposure to the dairy farm environment has a greater health significance for urban residents who have less prior exposure to VTEC and specific subgroups within the rural community. The latter is comprised of children with declining maternal immunity, the elderly, and other immunocompromised individuals who live on dairy farms and who can be considered an at risk group.

In 1993, a community outbreak (15 cases) of hemolytic uremic syndrome (HUS) in children occurred in a large area of Northern Italy over a period of several months and resulted in one death (Tozzi et al. 1994). No obvious epidemiologic link was observed among cases, except that they all lived in small townships. A case control study did not show an association between HUS and food or exposure to cattle, but suggested an association involving contact with chicken coops. However no VTEC were isolated from stool samples from the chicken coops. Live poultry has never been identified as a reservoir of VTEC (Beutin et al. 1993), even though *E. coli* O157:H7 has been isolated from retail poultry (Doyle et al. 1987). Detection of at least 3 different serogroups of VTEC (*E. coli* O157, *E. coli* O111, and *E. coli* O86) suggested multiple sources or vehicles of infection. The authors suggested living in rural settlements including exposure to live poultry was the risk factor, which should be considered in outbreak investigations. However, more powerful research is needed to confirm these observations.

Outbreaks of human enteric infection via direct contact with livestock have also been associated with cryptosporidiosis. In a study carried out by the Public Health Laboratory Service in England (CDR 1993), 23% of cryptosporidiosis cases were associated with farm animal contact. A significant association has been observed between cases of cryptosporidiosis in spring and contact with lambs. In England, a number of outbreaks of cryptosporidiosis have occurred among children visiting farms during the lambing seasons (CDR 1993; CDR 1994). Infected lambs that are scouring are not likely to be handled by visitors. However, lambs that are not scouring may excrete *Cryptosporidium* oocysts and could be an unrecognized reservoir of infection (CDR 1994).

Salmonella infection has also been associated with animal contact. In the spring of 1997, an outbreak of multidrug-resistant *Salmonella* Typhimurium DT104 struck a Vermont farm in Franklin County, USA (Spake 1997). It began as an epidemic among the cattle on the farm. By the end of the outbreak, 22 of the 147 cows had become ill, 13 had died, and 17 had become asymptomatic. Within days after the first death of a calf, a 5-year-old boy on the farm came down with the *Salmonella* infection. Soon after the first case, 7 other people were diagnosed. It was found that 90% of the farm family members who drank raw milk became ill. However milk was only one cause of the illness, caring for sick cattle was another. The last patient had cared for the sick cattle, but did not drink raw milk. She was the most serious patient and almost lost her life, due to the failure of some antibiotic treatments. The bacterium was resistant to penicillin and cephalosporin that were prescribed, in addition to being resistant to at least 5 other antibiotics (that are characteristic of the bacterium) including ampicillin, chloramphenicol, streptomycin, sulfonamides, and tetracycline. This was the first outbreak that conclusively demonstrated animal to human transmission of an antibiotic resistant bacterium in the U.S. In 1998, a ceftriaxone-resistant *Salmonella* infection was acquired by a child from cattle in western Nebraska (Fey et al. 2000). It was probable that the use of antimicrobial agents in cattle had led to the selection for the ceftriaxone-resistant strain that was subsequently transmitted to the child. In 1999, the Centers for Disease Control and Prevention (CDC) received reports from three state health departments (Idaho, Minnesota, and Washington) of outbreaks of multidrug-resistant *S.* Typhimurium infections in employees and clients of small animal veterinary clinics and an animal shelter (CDC 2001d). These outbreaks are the first to associate *S.* Typhimurium DT 104 with pets in the country. It was suggested that the inadvertent ingestion of animal feces or food contaminated with animal feces may have occurred as the result of suboptimal sanitation and hygienic practices in the veterinary facilities. Some of the cats in the facilities had a diarrheal illness that may have contributed to *Salmonella* transmission. Even after recovery from an acute bout of *Salmonella* gastroenteritis, fecal shedding of *Salmonella* can occur and may last for several months. In addition, the use of antimicrobial agents in veterinary facilities may have contributed to selection for multidrug-resistant *Salmonella* by lowering the infectious dose required to cause illness in animals and increasing the likelihood of transmission to humans. These outbreaks demonstrate that small animals besides cattle and lambs shed *Salmonella* and that small animal facilities in addition to farms, can also serve as a risk setting for transmission of *Salmonella* to animals and humans. In the U.S., multidrug-resistant *S.* Typhimurium DT 104 has been found in sheep, pigs, horses, goats, cats, dogs, elk, mice, coyotes, squirrels, raccoons, chipmunks, pigeons, and starlings. The birds are a real concern because their droppings can spread disease everywhere.

Farm workers are usually aware of health risks from zoonotic microorganisms, symptoms that indicate infection, and precautions that need to be taken. Farmers need to consider the health risks to visitors especially vulnerable groups, such as young children and pregnant women, and the precautions necessary to protect them. There have been some published guidelines that serve these purposes (Dawson et al. 1995). The public should be aware of farm visit-related risks and avoid hand to mouth contact when visiting farms. Health officials need to continue to educate the public and owners of farm visitor centers about disease that may be associated with farms.

Person-to-Person Transmission

Many enteric pathogens can be transmitted by person-to-person contact. They include *Campylobacter* spp., *Cryptosporidium*, *Escherichia coli* O157:H7, *Salmonella* spp., *Shigella* spp., hepatitis A, and Norwalk virus (Dawson et al. 1995; Becker et al. 2000; Health Canada 2000e; Health Canada 1997b; Health Canada 2001c; Jiang & Doyle 1999). In some cases, the disease is manifested as a secondary infection, obtained from a primary infection that was acquired through foodborne or waterborne routes. One such example is an outbreak of Norwalk virus which occurred during a college football game (Becker et al. 2000). During the game in Florida, diarrhea and vomiting developed in many of the members of a North Carolina team. The next day, similar symptoms developed in some of the players on the opposing team. The two football teams shared no food or beverages and had no contact off the playing field. The stool samples obtained from patients of both teams shared identical PCR sequences of a Norwalk virus. The outbreak investigation identified a turkey sandwich as the source of infection and it accounted for 95% of the primary cases. The football game spread the virus from the primary cases to other players. Both fecal-oral transmission and aerosol transmission of vomitus had probably occurred, given the intense physical contact and the use of bare hands during the game. It was advised that persons with acute gastroenteritis should be excluded from playing contact sports.

Hepatitis A virus is commonly transmitted via direct contact between people. In Canada, British Columbia has one of the highest rates of hepatitis A (Health Canada 2000e). In 1998, the province had a rate of 9.65 cases per 100,000, and men who had sex with men (MSM) and infected drug users (IDU) were identified as primary risk factors. In previous years cases involving Aboriginal populations contributed significantly to the provincial statistics. In a recent outbreak in August 1999, 14 cases were reported and a First Nations reserve was involved. Investigations identified household density as a contributing factor. Household density was often increased by houseguests who in turn propagated the outbreak. Over half of the cases in this outbreak occurred in children less than 14 years of age. Clinically, children are less likely to be symptomatic than adults and may shed viruses for extended periods of time. Poor hygiene and play activities may have contributed to the transmission of virus among children. The outbreak resulted in a decision to actively immunize the population of the reserve. In January 1997, a public health unit in southern Ontario was alerted to 4 cases of hepatitis A among a family, which belonged to a self- contained religious community (Health Canada 1997b). Over the course of 4 months, a total of 21 cases were identified within this community. In the index family, the mother was the first to become ill and during her illness, she had prepared food for large holiday season gatherings. The family outbreak of gastroenteritis illness led to a community outbreak. The mode of transmission was not confirmed, but both foodborne and person-to-person transmission had likely occurred. It was noted that children with unrecognized and unchecked subclinical illness can act as a reservoir for the disease for several months, placing older persons at continuous risk of clinical illness.

Shigellosis is transmitted by direct or indirect fecal-oral contact. Outbreaks are traditionally associated with conditions of crowding or where personal hygiene is poor, such as in prisons, day care centers, and psychiatric institutions. Outbreaks among men who have sex with men (MSM) are also common and are linked to high-risk sexual practices. In developed countries such as Canada and the United States, *Shigella sonnei* is the most common serotype and *S. dysenteriae* is the least common. Cases of shigellosis among MSM in these countries are typically associated with *S. flexneri*. However in March 2001, a cluster of *S. sonnei* infection was reported to the B. C. Centre for Disease Control and Prevention and MSM were involved (Health Canada 2001c). No common risk factors such as food or travel were noted among the cases, with the exception of high-risk sexual practices reported by the cases. This cluster followed a similar cluster of *S. sonnei* infection involving MSM in B. C. in November 2000 where no other common risk factors were identified. Transmission of enteric pathogens among MSM through high-risk sexual practices has long been documented. The low infectious dose of *Shigella* facilitates the transmission and increases the chance of outbreak in this population.

Person to person contact has been identified as a mode of transmission for *Escherichia coli* O157:H7. Money plays a role in transmission of bacteria. In a study by Jiang & Doyle (1999) who determined the survival of *E. coli* O157:H7 and *Salmonella* Enteritidis on coin surfaces, results indicated coins can serve as potential vehicles for transmitting both pathogens during general circulation of currency. *E. coli* O157:H7 and *S.* Enteritidis survived up to 11 days and 9 days, respectively, on the surface of U.S. pennies, nickels, dimes and quarters at 25°C. Upon contact with coin surfaces, bacterial cells are subject to drying and the bactericidal effect of copper. *S.* Enteritidis cells were more sensitive to drying than *E. coli* O157:H7. Survival of both pathogens was greatest on dimes and quarters and least on pennies, suggesting that the additional copper in the coin had enhanced bactericidal activity. This is of special significance to food handlers, who should use an intervention treatment such as washing hands after handling coins and before handling food.

CONCLUSIONS

Foodborne transmission is likely the most common route of transmission for enteric disease agents. Meat and poultry account for the greatest number of foodborne outbreaks. Other significant foods include produce, seafood, and dairy products. Enteric pathogens usually reside in the GI tract of food animals, either symptomatically or asymptomatically, and in the animal environment. There seems to be a continuous re-infection among animals and between them and the environment in which they reside. Most risks due to seafood consumption originate from the environment. Enteric pathogens associated with seafood come from natural marine or freshwater environments, fecal pollution of the natural environments, or processing and preparation environments. As a general rule, to reduce the risk of enteric infections, consumers should avoid eating raw or undercooked seafood particularly molluscan shellfish. Safety of oyster harvesting areas and aquaculture ponds relies on prevention of accidental contamination from human waste and intentional pollution to increase aquaculture production. Enteric infection associated with dairy foods usually arises from direct consumption of unpasteurized milk or its use as an ingredient in manufacture of dairy products. Dairy foods made from pasteurized milk can carry risk if improper pasteurization and post pasteurization contamination occur in processing plants. Fruit and vegetables are vulnerable to microbial contamination from the original production site to final presentation on the table. During production, produce can be contaminated with human enteric pathogens from soil, fertilizer, irrigation water, pesticide spray, wild and domestic animal waste, or field workers. Fresh produce that undergoes mechanical processing or minimal thermal treatment is especially prone to microbial propagation.

Waterborne disease has the potential to cause extensive outbreaks given the size of populations served by municipal distribution systems and the large number of people using recreational water facilities. Disease outbreaks associated with drinking water are a result of either poor protection of the source water or inadequate water treatment. Enteric illness due to recreational water activities has been on the rise in recent years. Sources of the pathogens in these waters often are the users themselves, in particular the young who are also the victims of the diseases. While proper disinfection remains the main control of disease in recreational water pools, elimination of fecal contamination through education and cooperation of the users and the general public is more effective in prevention of disease, particularly in non-chlorinated water such as lakes and ponds.

Airborne transmission of enteric disease is not frequently detected. This is due in part to our limited knowledge and use of unproven detection techniques for airborne pathogens. Major sources of airborne pathogens include human activity (such as sneezing, coughing, and talking), animal movement and waste, sewage and sewage treatment plants, and dust. Airborne disease is more apparent in farm, sewage plant, and other agriculture-related workers. This is also considered an occupational hazard.

Direct-contact transmission of enteric infection is usually a result of contact with infected animals or persons. Children's farms (petting zoos) are recognized as a risk setting for animal-person disease transmission. The dairy farm is another, particularly for children, the elderly, and the immunocompromised that live there and don't regularly come in contact with livestock. Person to person contact during sporting events, high-risk sexual practices, child play activities, elevated household densities, and general circulation of currency, can also spread enteric disease.

Identification of the sources of pathogens is of paramount importance in development of effective intervention strategies. This relies largely on the surveillance data from foodborne disease outbreaks and sporadic cases. Unfortunately in Canada, there is no one system that is used for consistently collecting foodborne disease data across the country. To date, national foodborne disease data are generated by a compilation of many different data sets, including those from provincial governments, the National Enterics Surveillance Program, the Enteric Disease Surveillance System, the Health of Animal Laboratories, the National Laboratory for Enteric Pathogens, and the Bureau of Infectious Disease (Cuff et al. 2000). These heterogeneous data sets pose difficulty in determining the real extent of foodborne disease problems. Because each data set records a different level of detail, subjective judgment is often required in the assignment of cause and management of data. In view of these inconsistencies and the probability of under-reporting, the need for a systematic, consistent data-collecting program is apparent.

Producers, processors, distributors, retailers, consumers, and regulators all have a role in minimizing human exposure to agents causing foodborne illness and enteric disease. Reduction and control of pathogen transmission at the farm level is feasible. This of course requires participation and involvement of farm owners and operators. Farmers' knowledge and attitude towards on-farm food safety programs are important for implementation of successful intervention strategies. Prevention of enteric disease is a constant challenge for the general public, health officials, and researchers, but it is not impossible.

REFERENCES

Ackman, D., S. Marks, P. Mack, M. Caldwell, T. Root, and G. Birkhead. 1997. Swimming-associated haemorrhagic colitis due to *Escherichia coli* O157:H7 infection: evidence of prolonged contamination of a fresh water lake. Epidemiol. Infect. 119:1–8.

Adams, A. P., and J. C. Spendlove. 1970. Coliform aerosols emitted by sewage treatment plants. Sci. 169:1218–1220.

Ahmed, F. E. 1992. Review: assessing and managing risk due to consumption of seafood contaminated with microorganisms, parasites, and natural toxins in the US. Int. J. Food Sci. Technol. 27:243–260.

Alamanos, Y., V. Maipa, S. Levidiotou, and E. Gessouli. 2000. A community waterborne outbreak of gastro-enteritis attributed to *Shigella sonnei*. Epidemiol. Infect. 125:499–503.

Al-Dagal, M., and D. Y. C. Fung. 1990. Aeromicrobiology- a review. Food Sci. Nutr. 29:333–340.

Al-Ghazali, M. R., and S. K. Al-Azawi. 1988. Effects of sewage treatment on the removal of *Listeria monocytogenes*. J. Appl. Bacteriol. 65:203–208.

Al-Ghazali, M. R., and S. K. Al-Azawi. 1990. *Listeria monocytogenes* contamination of crops grown on soil treated with sewage sludge cake. J. Appl. Bacteriol. 69:642–647.

Altekruse, S. F., M. L. Cohen, and D. L. Swerdlow. 1997. Emerging foodborne diseases. Emerg. Infect. Dis. 3:285–293.

Baloda, S. B., L. Christensen, and S. Trajcevska. 2001. Persistence of a *Salmonella enterica* serovar Typhimurium DT12 clone in a piggery and in agricultural soil amended with *Salmonella*-contaminated slurry. Appl. Environ. Microbiol. 67:2859–2862.

Barrell, R. A. E., P. R. Hunter, G. and Nichols. 2000. Microbiological standards for water and their relationship to health risk. Commun. Dis. Public Health, 3:8–13.

Becker, K. M., C. L. Moe, K. L. Southwick, and J. N. MacCormack. 2000. Transmission of Norwalk virus during a football game. N. Engl. J. Med. 343:1223–1227.

Bell, A., R. Guasparini, D. Meeds, R. G. Mathias, and J. D. Farley. 1993. A swimming pool associated outbreak of cryptosporidiosis in British Columbia. Can. J. Public Health, 84:334–337.

Besser, R. E., S. M. Lett, J. T. Weber, M. P. Doyle, T. J. Barett, J. G. Wells, and P. M. Griffin. 1993. An outbreak of diarrhea and hemolytic uremic syndrome from *Escherichia coli* O157:H7 in fresh-pressed apple cider. J. Am. Med. Assoc. 269:2217–2220.

Beuchat, L. R., and J-H. Ryu. 1997. Produce handling and processing practices. Emerg. Infect. Dis. 3:459–465.

Beutin, L., D. Geier, H. Steinruck, S. Zimmermann, and F. Scheutz. 1993. Prevalence and some properties of verotoxin (shiga-like toxin)-producing *Escherichia coli* in seven different species of healthy domestic animals. J. Clin. Microbiol. 31:2483–2488.

Bhaskar, N., and T. M.R. Setty. 1994. Incidence of vibrios of public health significance in the farming phase of tiger shrimp (*Penaeus monodon*). J. Sci. Food Agric. 66:225–231.

Blostein, J. 1991. Shigellosis from swimming in a park pond in Michigan. Public Health Rep. 106:317–322.

Brewster, D. H., M. I. Brown, D. Robertson, G. L. Houghton, J. Bimson, and J. C. M. Sharp. 1994. An outbreak of *Escherichia coli* associated with children's paddling pool. Epidemiol. Infect. 112:441–447.

Bryan, F. L., and M. P. Doyle. 1995. Health risks and consequences of *Salmonella* and *Campylobacter jejuni* in raw poultry. J. Food Protect. 58:326–344.

Cassin, M. H., A. M. Lammerding, E. C. D. Todd, W. Rose, and R. S. McColl. 1998. Quantitative risk assessment for *Escherichia coli* O157:H7 in ground beef hamburgers. Int. J. Food Microbiol. 41:21–44.

Center for Science in the Public Interest (CSPI). 1999. Produce-related outbreaks, 1990–1999 [Online]. URL: http://www.cspinet.org/new/prodhark.html (Accessed: February 18, 2002).

The Centers for Disease Control and Prevention (CDC). 1990. Epidemiologic notes and reports swimming-associated cryptosporidiosis-Los Angeles County. Morb. Mortal. Wkly. Rep. 39:343–345.

CDC. 1994a. Emerging infectious disease outbreak of *Salmonella* Enteritidis associated with nationally distributed ice cream products-Minnesota, South Dakota, and Wisconsin, 1994. Morb. Mortal. Wkly. Rep. 43:740–741.

CDC. 1994b. *Cryptosporidium* infections associated with swimming pools-Dane County, Wisconsin, 1993. Morb. Mortal. Wkly. Rep. 43:561–563.

CDC. 1996a. Outbreak of trichinellosis associated with eating cougar jerky – Idaho, 1995. Morb. Mortal. Wkly. Rep. 45:205–206.

CDC. 1996b. Salmonellosis associated with a thanksgiving dinner—Nevada, 1995. Morb. Mortal. Wkly. Rep. 45:1016–1017.

CDC. 1996c. *Shigella sonnei* outbreak associated with contaminated drinking water-Island Park, Idaho, August 1995. Morb. Mortal. Wkly. Rep. 45:229–231.

CDC. 1996d. Lake-associated outbreak of *Escherichia coli* O157:H7-Illinois, 1995. Morb. Mortal. Wkly. Rep. 45:437–439.

CDC. 1997a. Outbreak of staphylococcal food poisoning associated with precooked ham – Florida, 1997. Morb. Mortal. Wkly. Rep. 46:1189–1191.

CDC. 1997b. Viral gastroenteritis associated with eating oysters-Louisiana, December 1996–January 1997. Morb. Mortal. Wkly. Rep. 46:1109–1112.

CDC. 1998a. Outbreak of *Vibrio parahaemolyticus* infections associated with eating raw oysters-Pacific Northwest, 1997. Morb. Mortal. Wkly. Rep. 47:457–462.

CDC. 1998b. *Plesiomonas shigelloides* and *Salmonella* serotype Hartford infections associated with contaminated water supply-Livingston County, New York, 1996. Morb. Mortal. Wkly. Rep. 47:394–396.

CDC. 1999a. Outbreak of *Vibrio parahaemolyticus* infection associated with eating raw oysters and clams harvested from Long Island Sound-Connecticut, New Jersey, and New York, 1998. Morb. Mortal. Wkly. Rep. 48:48–51.

CDC. 1999b. Public health dispatch: outbreak of *Escherichia coli* O157:H7 and *Campylobacter* among attendees of the Washington County fair-New York, 1999. Morb. Mortal. Wkly. Rep. 48:803.

CDC. 2000. *Escherichia coli* O157:H7 [Online]. URL: http://www.cdc.gov/ncidod/dbmd/diseaseinfo/escherichiacoli_t.htm (Accessed: March 19, 2002).

CDC. 2001a. Preliminary FoodNet data on the incidence of foodborne illnesses—selected sites, United States, 2000. Morb. Mortal. Wkly. Rep. 50:241–246.

CDC. 2001b. Prevalence of parasites in fecal material from chlorinated swimming pools-United States, 1999. Morb. Mortal. Wkly. Rep. 50:410–412.

CDC. 2001c. Outbreaks of *Escherichia coli* O157:H7 infections among children associated with farm visits-Pennsylvania and Washington, 2000. Morb. Mortal. Wkly. Rep. 50:293–297.

CDC. 2001d. Outbreaks of multidrug-resistant *Salmonella* Typhimurium associated with veterinary facilities-Idaho, Minnesota, and Washington, 1999. Morb. Mortal. Wkly. Rep. 50:701–704.

CDC. 2002. Outbreak of *Salmonella* serotype *kottbus* infections associated with eating alfalfa sprouts-Arizona, California, Colorado, and New Mexico, February-April 2001. Morb. Mortal. Wkly. Rep. 51:7–9.

Chapman, P. A., J. Cornell, and C. Green. 2000. Infection with verocytotoxin-producing *Escherichia coli* O157 during a visit to an inner city open farm. Epidemiol. Infect. 125:531–536.

Chapman, P. A., C. A. Siddons, A. T. Cerdan Malo, and M. A. Harkin. 1997. A 1-year study of *Escherichia coli* O157:H7 in cattle, sheep, pigs, and poultry. Epidemiol. Infect. 119:245–250.

Cieslak, P. R., T. J. Barett, P. M. Griffin, K. F. Gensheimer, G. Beckett, J. Buffington, and M. G. Smith. 1993. *Escherichia coli* O157:H7 infection from a manured garden. Lancet, 342:367.

Coia, J. E., J. C. M. Sharp, D. M. Campbell, J. Curnow, and C. N. Ramsay. 1998. Environmental risk factors for sporadic *Escherichia coli* O157:H7 infection in Scotland: results of a descriptive epidemiology study. J. Infect. 36:317–321.

Collins, J. E. 1997. Impact of changing consumer lifestyles on the emergence/reemergence of food-borne pathogens. Emerg. Infect. Dis. 3:471–479.

Communicable Disease Report weekly (CDR). 1993. Cryptosporidiosis. Public Health Lab. Service, 3:89.

CDR. 1994. Cryptosporidiosis associated with farm visits. Public Health Lab. Service, 4:73.

Crampin, M., G. Willshaw, R. Hancock, T. Djuretic, C. Elstob, A. Rouse, T. Cheasty, and J. Stuart. 1999. Outbreak of *Escherichia coli* O157 infection associated with a music festival. Eur. J. Clin. Microbiol. Infect. Dis. 18:286–288.

Croonenberghs, R. E. 2000. Contamination of shellfish-growing area. In: R. E. Martin, E. P. Carter, G. J. Flick, Jr., and L. M. Davis (eds.). Marine & freshwater products handbook. Technomic publishing co., Inc., Lancaster, Pennsylvania. pp. 665–693.

Cuff, W. R., R. Ahmed, D. L. Woodward, C. G. Clark, and F. G. Rodgers. 2000. Enteric pathogens identified in Canada—annual summary 1998. National Laboratory for Enteric Pathogens, pp. 1, 67.

Current, W. L., and L. S. Garcia. 1991. Cryptosporidiosis. Clin. Microbiol. Rev. 4:325–358.

Dalsgaard, A. 1998. The occurrence of human pathogenic V*ibrio* spp. and *Salmonella* in aquaculture. Int. J. Food Sci. Technol. 33:127–138.

D'Aoust, J-Y., D. W. Warburton, and A. M. Sewell. 1985. *Salmonella typhimurium* phage-type 10 from cheddar cheese implicated in a major Canadian foodborne outbreak. J. Food Protect. 48:1062–1066.

Davis, L. J., H. L. Roberts, D. D. Juranek, S. R. Framn, and R. Soave. 1998. A survey of risk factors for cryptosporidiosis in New York City: drinking water and other exposures. Epidemiol. Infect. 121:357–367.

Dawson, A., R. Griffin, A. Fleetwood, and N. J. Barrett. 1995. Farm visits and zoonoses. Commun. Dis. Rep. 5:R81–R85.

DePaola, A., C. A. Kaysner, J. Bowers, and D. W. Cook. 2000. Environmental investigations of *Vibrio parahaemolyticus* in oysters after outbreaks in Washington, Texas, and New York (1997 and 1998). Appl. Environ. Microbiol. 66:4649–4654.

DeRegnier, D., L. Cole, D. G. Schupp, and S. Erlandsen. 1989. Viability of *Giardia* cysts suspended in lake, river, and tap water. Appl. Environ. Microbiol. 55:1223–1229.

DesRosiers, A., J. M. Fairbrother, R. P. Johnson, C. Desautels, A. Letellier, and S. Quessy. 2001. Phenotypic and genotypic characterization of *Escherichia coli* verotoxin-producing isolates from humans and pigs. J. Food Protect. 64:1904–1911.

Dev, V. J., M. Main, and I. Gould. 1991. Waterborne outbreak of *Escherichia coli* O157:H7. Lancet, 337:1412.

Doores, S. 1999. Food safety—current status and future needs. A report from the American Academy of Microbiology, pp. 7–14.

El-Gazzar, F. E., and E. H. Marth. 1992. Salmonellae, salmonellosis, and dairy foods: a review. J. Dairy Sci. 75:2327–2343.

Ellis, A., M. Preston, A. Borczyk, B. Miller, P. Stone, B. Hatton, A. Chagla, and J. Hockin. 1998. A community outbreak of *Salmonella berta* associated with a soft cheese product. Epidemiol. Infect. 120:29–35.

Fernandes. C. F., G. J. Flick, Jr, J. L. Silva, and T. A. McCaskey. 1997. Comparison of quality in aquacultured fresh catfish fillets II. Pathogens *E. coli* O157:H7, *Campylobacter*, *Vibrio*, *Plesiomonas*, and *Klebsiella*. J. Food Protect. 60:1182–1188.

Fey, P. D., T. J. Safranek, M. E. Rupp, E. F. Dunne, E. Ribot, P. C. Iwen, P. A. Bradford, F. J. Angulo, and S. H. Hinrichs. 2000. Ceftriaxone-resistant *Salmonella* infection acquired by a child from cattle. N. Engl. J. Med. 342:1242–1249.

Food and Drug Administration (FDA). 1998. Guidance to industry-guide to minimize microbial food safety hazards for fresh fruits and vegetables. U.S. Department of Health and Human Services, Center for Food Safety and Applied Nutrition, October 1998.

Food Safety and Inspection Service (FSIS). 1996. Nationwide pork microbiological baseline data collection program: market hogs [Online]. URL: http://www.fsis.usda.gov/OPHS/baseline/markhog1.pdf and http://www.fsis.usda.gov/OPHS/baseline/markhog2.pdf (Accessed: April 11, 2002).

FSIS. 2001. Progress report on Salmonella testing of raw meat and poultry products, 1998–2000 [Online]. URL: http://www.fsis.usda.gov/ophs/haccp/salmdata2.htm (Accessed: April 11, 2002).

Frenzen, P. D., E. E. DeBess, K. E. Hechemy, H. Kassenborg, M. Kennedy, K. McCombs, A. McNees, and the Foodnet Working Group. 2001. Consumer acceptance of irradiated meat and poultry in the United States. J. Food Protect. 64:2020–2026.

Furtado, C., G. K. Adak, J. M. Stuart, P. G. Wall, H. S. Evans, and D. P. Casemore. 1998. Outbreaks of waterborne infectious intestinal disease in England and Wales, 1992–5. Epidemiol. Infect. 121:109–119.

Garett, E. S., C. L. dos Santos, and M. L. Jahncke. 1997. Public, animal, and environmental health implications of aquaculture. Emerg. Infect. Dis. 3:453–457.

Gast, R. K., B. W. Mitchell, and P. S. Holt. 1999. Application of negative air ionization for reducing experimental airborne transmission of *Salmonella enteritidis* to chicks. Poult. Sci. 78:57–61.

Gaulin, C. D., D. Ramsay, P. Cardinal, M-A. D'Halevyn. 1999. Gastroenteritis outbreak of viral origin related to imported strawberries consumption. Can. J. Public Health, 90:37–40.

Greensmith, C. T., R. S. Stanwick, B. E. Elliot, and M. V. Fast. 1988. Giardiasis associated with the use of a water slide. Pediatr. Infect. Dis. J. 7:91–94.

Griffiths, M. W. 2000. The new face of food-borne illness. CMSA News, Can. Meat Sci. Assoc., Ottawa ON, March:6–9.

Guan, T. Y., G. Blank, A. Ismond, and R. V. Acker. 2001. Fate of foodborne bacterial pathogens in pesticide products. J. Sci. Food Agric. 81:503–512.

Hancock, D., T. Besser, J. Lejeune, M. Davis, and D. Rice. 2001. The control of VTEC in the animal reservoir. Int. J. Food Microbiol. 66:71–78.

Harter, L., F. Frost, G. Grunenfelder, K. Perkins-Jones, and J. Libby. 1984. Giardiasis in an infant and toddler swim class. Am. J. Public Health, 74:155–156.

Health Canada. 1995. Outbreak of *Campylobacter* infection among farm workers: an occupational hazard. Can. Commun. Dis. Rep. 21:153–156.

Health Canada. 1997a. Outbreak of *Vibrio parahaemolyticus* related to raw oysters in British Columbia. Can. Commun. Dis. Rep. 23:145–148.

Health Canada. 1997b. Hepatitis A outbreak in a socially-contained religious community in rural southern Ontario. Can. Commun. Dis. Rep. 23:161–166.

Health Canada. 1998. The distribution of foodborne disease by risk setting – Ontario. Can. Commun. Dis. Rep. 24:61–64.

Health Canada. 1999. Guidelines for raw ground beef products found positive for *Escherichia coli* O157:H7. Raw Foods of Animal Origin and Steering Committee. Guideline no. 10. Health Protection Branch, pp. 1–2.

Health Canada. 2000a. Interim guidelines for the control of verotoxinogenic *Escherichia coli* including *E. coli* O157:H7 in ready to eat fermented sausages containing beef or a beef product as an ingredient. Raw Foods of Animal Origin and Steering Committee. Guideline no. 12. Health Protection Branch, pp. 1–7.

Health Canada. 2000b. Case-control study assessing the association between yersiniosis and exposure to salami. Can. Commun. Dis. Rep. 26:161–164.

Health Canada. 2000c. Waterborne outbreak of gastroenteritis associated with a contaminated municipal water supply, Walkerton, Ontario, May–June 2000. Can. Commun. Dis. Rep. 26:170–173.

Health Canada. 2000d. Drinking water quality and health-care utilization for gastrointestinal illness in greater Vancouver. Can. Commun. Dis. Rep. 26:211–214.

Health Canada. 2000e. Hepatitis A in the northern interior of British Columbia: an outbreak among members of a First Nations community. Can. Commun. Dis. Rep. 26:157–161.

Health Canada. 2001a. Outbreak of trichinellosis associated with arctic walruses in Northern Canada, 1999. Can. Commun. Dis. Rep. 27:31–36.

Health Canada. 2001b. Waterborne cryptosporidiosis outbreak, North Battleford, Saskatchewan, spring 2001. Can. Commun. Dis. Rep. 27:185–192.

Health Canada. 2001c. Clusters of *Shigella sonnei* in men who have sex with men, British Columbia, 2001. Can. Commun. Dis. Rep. 27:109–114.

Health Canada. 2002a. *Escherichia coli* O157 outbreak associated with the ingestion of unpasteurized goat's milk in British Columbia, 2001. Can. Commun. Dis. Rep. 28:6–8.

Health Canada. 2002b. Managing the health risks associated with *Listeria monocytogenes* in ready-to-eat foods. Food Directorate, Health Products and Food Branch, Health Canada. Policy ID: 2002-FD-01. Date issued: January 8, 2002.

Health Canada. 2002c. Abbott's choice brand cheese products may contain *Listeria monocytogenes* [Online]. Health Hazard Alert, February 13, 2002. URL: http://www.inspection.gc.ca/english/corpaffr/recarapp/2002/20020213be.shtml (Accessed: February 15, 2002).

Hedberg, C. W., J. A. Korlath, J. Y. D'Aoust, K. E. White, W. L. Schell, M. R. Miller, D. N. Camero, K. L. MacDonald, and M. T. Osterholm. 1992. A multistate outbreak of *Salmonella javiana* and *Salmonella oranienburg* infections due to consumption of contaminated cheese. J. Am. Med. Assoc. 268:3203–3207.

Heng, B. H., K. T. Goh, S. Doraisingham, and G. H. Quek. 1994. Prevalence of hepatitis A virus infection among sewage workers in Singapore. Epidemiol. Infect. 113:121–128.

Henry, B., B. Warshawsky, I. Gutmanis, J. Reffle, J. Dow, G. Pollett, C. LeBer, M. Naus, F. Jamieson, R. Ahmed, and D. H. Werker. 2001. Outbreak of *Escherichia coli* O157:H7 infections associated with a petting zoo at a fall fair, Ontario, Canada, 1999 [Online]. Field Epidemiology Training Program 2001 Abstracts, Health Canada. URL: http://www.hc-sc.gc.ca/pphb-dgspsp/fetp-pfei/abs01_e.html (Accessed: December 15, 2001).

Hildebrand, J. M., H. C. Maguire, R. E. Holliman, and E. Kangesu. 1996. An outbreak of *Escherichia coli* O157:H7 infection linked to paddling pools. Commun. Dis. Rep. 6:R33–R36.

Ho, M-S., R. I. Glass, S. S. Monroe, H. P. Madore, S. Stine, P. F. Pinsky, D. Cubitt, C. Ashley, and E. O. Caul. 1989. Viral gastroenteritis aboard a cruise ship. Lancet, 2: 961–964.

Hume, M. E., D. J. Nisbet, S. A. Buckley, R. L. Ziprin, R. C. Anderson, and L. H. Stanker. 2001. Inhibition of in vitro *Salmonella typhimurium* colonization in porcine cecal bacteria continuous-flow competitive exclusion cultures. J. Food Protect. 64:17–22.

Isaac-Renton, J., W. Moorehead, and A. Ross. 1996. Longitudinal studies of *Giardia* contamination in two community drinking water supplies: cyst levels, parasite viability, and health impact. Appl. Environ. Microbiol. 62:47–54.

Isaacson, M., P. H. Canter, P. Effler, L. Arntzen, P. Bomans, and R. Heenan. 1993. Haemorrhagic colitis epidemic in Africa. Lancet, 341:961.

Jackson, S. G., R. B. Goodbrand, R. P. Johnson, V. G. Odorico, D. Alves, K. Rahn, J. B. Wilson, M. K. Welch, and R. Khakhria. 1998. *Escherichia coli* O157:H7 diarrhoea associated with well water and infected cattle on an Ontario farm. Epidemiol. Infect. 120:17–20.

Jiang, X., and M. P. Doyle. 1999. Fate of *Escherichia coli* Oi157:H7 and *Salmonella* Enteritidis on currency. J. Food Protect. 62:805–807.

Joce, R. E., J. Bruce, D. Kiely, N. D. Noah, W. B. Dempster, R. Stalker, P. Gumsley, P. A. Chapman, P. Norman, J. Watkins, H. V. Smith, T. J. Price, and D. Watts. 1991. An outbreak of cryptosporidiosis associated with a swimming pool. Epidemiol. Infect. 107:497–508.

Jones, D. L. 1999. Potential health risks associated with the persistence of *Escherichia coli* O157 in agricultural environments. Soil Use Manage. 15:76–83.

Jones, K. 2000. Seven dead from e-coli contamination in Ontario, Canada [Online]. URL: http://www.wsws.org (Accessed: April 9, 2002).

Karmali, M. A., B. T. Steele, M. Petric, and C. Lim. 1983. Sporadic cases of hemolytic-uremic syndrome associated with faecal cytotoxin and cytotoxin producing *Escherichia coli* in stools. Lancet, 1:619–620.

Keene, W. E., J. M. McAnulty, F. C. Hoesly, L. P. Williams, K. Hedberg, G. L. Oxman, T. J. Barrett, M. A. Pfaller, and D. W. Fleming. 1994. A swimming-associated outbreak of hemorrhagic colitis caused by *Escherichia coli* O157:H7 and *Shigella sonnei*. N. Engl. J. Med. 331:579–584.

Keene, W. E., E. Sazie, J. Kok, D. H. Rice, D. D. Hancock, V. K. Balan, T. Zhao, and M. P. Doyle. 1997. An outbreak of *Escherichia coli* O157:H7 infections traced to jerky made from deer meat. J. Am. Med. Assoc. 277:1229–1231.

Kotula, A. W., and B. S. Emswiler-Rose. 1988. Airborne microorganisms in a pork processing establishment. J. Food Protect. 51:935–937.

Kramer, J. M., J. A. Frost, F. J. Bolton, and D. R. A. Wareing. 2000. *Campylobacter* contamination of raw meat and poultry at retail sale: identification of multiple types and comparison with isolates from human infection. J. Food Protect. 63:1654–1659.

Laberge, I., M. W. Griffiths, and M. W. Griffiths. 1996. Prevalence, detection and control of *Cryptosporidium parvum* in food. Int. J. Food Microbiol. 31:1–26.

Licence, K., K. R. Oates, B. A. Synge, and T. M. S. Reid. 2001. An outbreak of *E. coli* O157 infection with evidence of spread from animals to man through contamination of a private water supply. Epidemiol. Infect. 126:135–138.

Lindell, S. S., and P. Quinn. 1973. *Shigella sonnei* isolated from well water. Appl. Microbiol. 26:424–425.

Lindsay J. A. 1997. Chronic sequelae of foodborne disease. Emerg. Infect. Dis. 3:443–452.

Mac Kenzie, W. R., N. J. Hoxie, M. E. Proctor, M. S. Gradus, K. A. Blair, D. E. Peterson, J. J. Kazmierczak, D. G. Addiss, K. R. Fox, J. B. Rose, and J. P. Davis. 1994. A massive outbreak in Milwaukee of *Cryptosporidium* infection transmitted through the public water supply. N. Engl. J. Med. 331:161–167.

Mackey, B. 1989. The incidence of food poisoning bacteria on red meat and poultry in the United Kingdom. Food Sci. Technol. Today, 3:246–250.

MacLean, J. D., J. Viallet, C. Law, and M. Staudt. 1989. Trichinosis in the Canadian Arctic: report of five outbreaks and a new clinical syndrome. J. Infect. Dis. 160:513–520.

Madden, J. M. 1992. Microbial pathogens in fresh produce-the regulatory perspective. J. Food Protect. 55:821–823.

Mahoney, F. J., T. A. Farley, K. Y. Kelso, S. A. Wilson, J. M. Horan, and L. M. McFarland. 1992. An outbreak of hepatitis A associated with swimming in a public pool. J. Infect. Dis. 165:613–618.

Marshall, B., P. Flynn, D. Kamely, and S. B. Levy. 1988. Survival of *Escherichia coli* with and without ColE1::Tn5 after aerosol dispersal in a laboratory and a farm environment. Appl. Environ. Microbiol. 54:1776–1783.

Martin, J. H., and D. L. Marshall. 1995. Characteristics and control of potential foodborne pathogens in cultured dairy foods. Cultured Dairy Products J. 30:9–16.

McAnulty, J. M., D. W. Fleming, and A. H. Gonzalez. 1994. A community-wide outbreak of cryptosporidiosis associated with swimming at a wave pool. J. Am. Med. Assoc. 272:1597–1600.

McAnulty, J. M., W. E. Keene, D. Leland, F. Hoesly, B. Hinds, G. Stevens, and D. W. Fleming. 2000. Contaminated drinking water in one town manifesting as an outbreak of cryptosporidiosis in another. Epidemiol. Infect. 125:79–86.

McGowan, K. L., Wickersham, E., and N. A. Strockbine. 1989. *Escherichia coli* O157:H7 from water. Lancet, i:967–968.

Mead P. S., L. Slutsker, V. Dietz, L. F. McCaig, J. S. Bresee, C. Shapiro, P. M. Griffin, and R. V. Tauxe. 1999. Food-related illness and death in the United States. Emerg. Infect. Dis. 5:607–625.

Millard, P. S., K. F. Gensheimer, D. G. Addiss, D. M. Sosin, G. A. Beckett, A. Houck-Jankoski, and A. Hudson. 1994. An outbreak of cryptosporidiosis from fresh-pressed apple cider. J. Am. Med. Assoc. 272:1592–1596.

Milne, L. M., A. Pom, I. Strudley, G. C. Pritchard, R. Crooks, M. Hall, G. Duckworth, C. Seng, M. D. Susman, J. Kearney, R. J. Wiggins, M. Moulsdale, T. Cheasty, and G. A. Willshaw. 1999. *Escherichia coli* O157:H7 incident associated with a farm open to members of the public. Commun. Dis. Public Health, 2:22–26.

Modi, R., Y. Hirvi, A. Hill, and M. W. Griffiths. 2001. Effect of phage on survival of *Salmonella* Enteritidis during manufacture and storage of cheddar cheese made from raw and pasteurized milk. J. Food Protect. 64:927–933.

Morgan, D., C. P. Newman, D. N. Hutchinson, A. M. Walker, B. Rowe, and F. Majid. 1993. Verotoxin producing *Escherichia coli* 157 infections associated with the consumption of yoghurt. Epidemiol. Infect. 111:181–187.

Morgan, G. M., C. Newman, S. R. Palmer, J. B. Allen, W. Shepherd, A. M. Rampling, R. E. Warren, R. J. Gross, S. M. Scotland, and H. R. Smith. 1988. First recognized community outbreak of haemorrhagic colitis due to verotoxin producing *Escherichia coli* O157:H7 in the UK. Epidemiol. Infect. 101:83–91.

Nijsten, R., N. London, A. Van Den Bogaard, and E. Stobberingh. 1994. Resistance in faecal *Escherichia coli* isolated from pigfarmers and abattoir workers. Epidemiol. Infect. 113:45–52.

O'Brien, S. J., G. K. Adak, and C. Gilham. 2001. Contact with farming environment as a major risk factor for Shiga toxin (vero cytotoxin)-producing *Escherichia coli* O157:H7 infection in humans. Emerg. Infect. Dis. 7:1049–1051.

Ogden, I. D., D. R. Fenlon, A. J. A. Vinten, and D. Lewis. 2001. The fate of *Escherichia coli* O157:H7 in soil and its potential to contaminate drinking water. Int. J. Food Microbiol. 66:111–117.

Ostroff, S. M., G. Kapperud, L. C. Hutwagner, T. Nesbakken, N. H. Bean, J. Lassen, and R. V. Tauxe. 1994. Sources of sporadic *Yersinia enterocolitica* infections in Norway: a prospective case-control study. Epidemiol. Infect. 112:133–141.

Paunio, M., R. Pebody, M. Keskimaki, M. Kokki, P. Ruutu, S. Oinonen, V. Vuotari, A. Siitonen, E. Lahti, and P. Leinikki. 1999. Swimming-associated outbreak of *Escherichia coli* O157:H7. Epidemiol. Infect. 122:1–5.

Payment, P. 1989. Bacterial colonization of domestic reverse-osmosis water filtration units. Can. J. Microbiol. 35:1065–1067.

Payment, P., A. Berte, M. Prevost, B. Menard, and B. Barbeau. 2000. Occurrence of pathogenic microorganisms in the Saint Lawrence River (Canada) and comparison of health risks for populations using it as their source of drinking water. Can. J. Microbiol. 46:565–576.

Payment, P., A. Berube, D. Perreault, R. Armon, and M. Trudel. 1989. Concentrations of *Giardia lamblia* cysts, *Legionella pneumophila*, *Clostridium perfringens*, human enteric viruses, and coliphages from large volumes of drinking water, using a single filtration. Can. J. Microbiol. 35:932–935.

Pearson, A. D., M. Greenwood, T. D. Healing, D. Rollins, M. Shahamat, J. Donaldson, and R. R. Colwell. 1993. Colonization of broiler chickens by waterborne *Campylobacter jejuni*. Appl. Environ. Microbiol. 59:987–996.

Pillai, S. D., K. W. Widmer, S. E. Dowd, and S. C. Ricke. 1996. Occurrence of airborne bacteria and pathogen indicators during land application of sewage sludge. Appl. Environ. Microbiol. 62:296–299.

Pontello, M., L. Sodano, A. Nastasi, C. Mammina, M. Astuti, M. Domenichini, G. Belluzzi, E. Soccini, M. G. Silvestri, M. Gatti, E. Gerosa, and A. Montagna. 1998. A community-based outbreak of *Salmonella enterica* serotype Typhimurium associated with salami consumption in Northern Italy. Epidemiol. Infect. 120:209–214.

Porter, J. D., H. P. Ragazzoni, J. D. Buchanon, H. A. Waskin, D. D. Juranek, and W. E. Parkin. 1988. *Giardia* transmission in a swimming pool. Am. J. Public Health, 78:659–662.

Potter, A., S. Gomis, G. Mutwiri, and D. Wilson. 2000. Vaccines for the prevention of *Escherichia coli* O157:H7 colonization of cattle [Online]. Inventory Can. Agri-Food Res. URL: http://res1.agr.ca/pls/icarweb/icarqueryeng.display?p_year=2001&p_icar_id=33330408 (Accessed: January 21, 2002).

Puech, M. C., J. M. McAnulty, M. Lesjak, N. Shaw, L. Heron, and J. M. Watson. 2001. A statewide outbreak of cryptosporidiosis in New South Wales associated with swimming at public school. Epidemiol. Infect. 126:389–396.

Rabsch, W., B. M. Hargis, R. M. Tsolis, R. A. Kingsley, K-H. Hinz, H. Tschäpe, and A. J. Bäumler. 2000. Competitive exclusion of *Salmonella* Enteritidis by *Salmonella* Gallinarum in poultry. Emerg. Infect. Dis. 6:443–448.

Ratnam, S., F. Stratton, C. O'Keefe, A. Roberts, R. Coates, M. Yetman, S. Squires, R. Khakhria, and J. Hockin. 1999. *Salmonella enteritidis* outbreak due to contaminated cheese-Newfoundland. Can. Commun. Dis. Rep. 25:17.

Robertson, L. J., A. T. Campbell, and H. V. Smith. 1992. Survival of *Cryptosporidium parvum* oocysts under various environmental pressures. Appl. Environ. Microbiol. 55:1519–1522.

Rose, J. B., and T. R. Slifko. 1999. *Giardia Cryptosporidium*, and *Cyclospora* and their impact on foods: a review. J. Food Protect. 62:1059–1070.

Satterthwaite, P., K. Pritchard, and D. Floyd. 1999. Case-control study of *Yersinia* infections in Auckland. Aust. N. Z. J. Public Health, 23:482–485.

Schlech, W. F., P. M. Lavigne, R. A. Bortolussi, A. C. Allen, E. V. Haldene, A. J. Wort, A. W. Hightower, S. E. Johnston, S. H. King, E. S. Nicholls, and C. V. Broome. 1983. Epidemic listeriosis-evidence for transmission by food. N. Engl. J. Med. 308:203–206.

Seo, K. H., B. W. Mitchell, P. S. Holt, and R. K. Gast. 2001. Bactericidal effects of negative air ions on airborne and surface *Salmonella* Enteritidis from an artificially generated aerosol. J. Food Protect. 64:113–116.

Shiferaw, B., S. Yang, P. Cieslak, D. Vugia, R. Marcus, J. Koehler, V. Deneen, F. Angulo, and the FoodNet working group. 2000. Prevalence of high-risk food consumption and food-handling practices among adults: a multistate survey, 1996 to 1997. J. Food Protect. 63:1538–1543.

Shukla, R., R. Slack, A. George, T. Cheasty, B. Rowe, and J. Scutter. 1995. *Escherichia coli* O157 infection associated with a farm visitor center. Commun. Dis. Rep. 5:R86–R90.

Smith, H. V., C. A. Paton, M. M. A. Mitambo, and R. W. A. Girdwood. 1997. Sporulation of *Cyclospora* sp. oocysts. Appl. Environ. Microbiol. 63:1631–1632.

Solomon, E. B., S. Yaron, and K. R. Matthews. 2002. Transmission of *Escherichia coli* O157:H7 from contaminated manure and irrigation water to lettuce plant tissue and its subsequent internalization. Appl. Environ. Microbiol. 68:367–400.

Sorvillo, F. J., S. H. Waterman, J. K. Vogt, and N. England. 1988. Shigellosis associated with recreational water contact in Los Angeles County. Am. J. Tropic. Med. Hyg. 38:613–617.

Spake, A. 1997. O is for outbreak. U.S. News World Rep. 123:70–84.

Steele, B. T., N. Murphy, and C. P. Rance. 1982. An outbreak of hemolytic uremic syndrome associated with ingestion of fresh apple juice. J. Pediatr. 101:963–965.

Swanenburg, M., H. A. P. Urlings, D. A. Keuzenkamp, and J. M. A. Snijders. 2001. *Salmonella* in the lairage of pig slaughterhouses. J. Food Protect. 64:12–16.

Swerdlow, D. L., B. A. Woodruff, R. C. Brady, P. M. Griffin, S. Tippen, H. D. Donnell, Jr., E. Geldreich, B. J. Payne, A. Meyer, Jr., J. G. Wells, et al. 1992. A waterborne outbreak in Missouri of *Escherichia coli* O157:H7 associated with bloody diarrhea and death. Ann. Int. Med. 117:812–819.

Tauxe, R. V. 1997. Emerging foodborne diseases: an evolving public health challenge. Emerg. Infect. Dis. 3:425–434.

Tauxe, R., H. Kruse, C. Hedberg, M. Potter, J. Madden, and K. Wachsmuth. 1997. Microbial hazards and emerging issues associated with produce-a preliminary report to the National Advisory Committee on Microbiologic Criteria for Foods, 60:1400–1408.

Tauxe, R. V., J. Vandepitte, G. Wauters, S. M. Martin, V. Goossens, P. De Mol, R. Van Noyen, and G. Thiers. 1987. *Yersinia enterocolitica* infections and pork: the missing link. Lancet, 1:1129–1132.

Teltsch, B., and E. Katzenelson. 1978. Airborne enteric bacteria and viruses from spray irrigation with wastewater. Appl. Environ. Microbiol. 35:290–296.

Thorne, P. S., M. S. Kiekhaefer, P. Whitten, and K. J. Donham. 1992. Comparison of bioaerosol sampling methods in barns housing swine. Appl. Environ. Microbiol. 58:2543–2551.

Tillett, H. E., J. de Louvois, and P. G. Wall. 1998. Surveillance of outbreaks of waterborne infectious disease: categorizing levels of evidence. Epidemiol. Infect. 120:37–42.

Todd, E. C. D.; P. Chatman, and V. Rodrigues. 2000. Annual summaries of foodborne and water-borne disease in Canada, 1994 and 1995, Health Products and Food Branch, Health Canada, Polyscience Publications Inc., Laval, Quebec.

Tozzi, A. E., A. Niccolini, A. Caprioli, I. Luzzi, G. Montini, G. Zacchello, A. Gianviti, F. Principato, and G. Rizzoni. 1994. A community outbreak of haemolytic-uraemic syndrome in children occurring in a large area of Northern Italy over a period of several months. Epidemiol. Infect. 113:209–219.

Tschäpe, H., R. Prager, W. Streckel, A. Fruth., E. Tietze, and G. Böhme. 1995. Verotoxinogenic *Citrobacter freundii* associated with severe gastroenteritis and cases of haemolytic uraemic syndrome in a nursery school: green butter as the infection source. Epidemiol. Infect. 114:441–450.

University of Guelph press release (UGPR). 1997. *E. coli*-contaminated well water linked to gastrointestinal illness. May 22.

Van Donkersgoed, J., T. Graham, and V. Gannon. 1999. The prevalence of verotoxins, *Escherichia coli* O157:H7, and *Salmonella* in the feces and rumen of cattle at processing. Can. Vet. J. 40:332–338.

Vasavada, P. C. 1988. Pathogenic bacteria in milk—a review. J. Dairy Sci. 71:2809–2816.

Wachtel, M. R., L. C. Whitehand, and R. E. Mandrell. 2002. Association of *Escherichia coli* O157:H7 with preharvest leaf lettuce upon exposure to contaminated irrigation water. J. Food Protect. 65:18–25.

Wallis, P. M., S. L. Erlandsen, J. L. Isaac-Renton, M. E. Olson, W. J. Robertson, and H. Van Keulen. 1996. Prevalence of *Giardia* cysts and *Cryptosporidium* oocysts and characterization of *Giardia* spp. isolated from drinking water in Canada. Appl. Environ. Microbiol. 62:2789–2797.

Wang, G., T. Zhao, and M. P. Doyle. 1996. Fate of Enterohemorrhagic *Escherichia coli* O157:H7 in bovine feces. Appl. Environ. Microbiol. 62:2567–2570.

Warburton, D. W. 1993. A review of the microbiological quality of bottled water sold in Canada. Part 2. The need for more stringent standards and regulations. Can. J. Microbiol. 39:158–168.

Warburton, D. W., and K. L. Dodds. 1992. A review of the microbiological quality of bottled water sold in Canada between 1981 and 1989. Can. J. Microbiol. 38:12–19.

Warshawsky, B. 2001. An *E. coli* O157:H7 outbreak associated with an animal exhibit [Online]. Middlesez-London Health Unit Investigation and Recommendations. Executive Summary. URL: http://www.healthunit.com/template.asp?id=859 (Accessed: March 4, 2002).

Wells, J. G., L. D. Shipman, K. D. Greene, E. G. Sowers, J. H. Green, D. N. Cameron, F. P. Downes, M. L. Martin, M. Griffin, S. M. Ostroff, M. E. Potter, R. V. Tauxe, and I. K. Wachsmuth. 1991. Isolation of *Escherichia coli* serotype O157:H7 and other shiga-like-toxin-producing *E. coli* from dairy cattle. J. Clin. Microbiol. 29:985–989.

Wells, S. J., P. J. Fedorka-Cray, D. A. Dargatz, K. Ferris, and A. Green. 2001. Fecal shedding of *Salmonella* spp. by dairy cows on farm and at cull cow markets. J. Food Protect. 64:3–11.

Wilson, J., J. Spika, R. Clarke, R. Johnson, S. Renwick, M. Karmali, H. Lior, D. Alves, and C. Gyles. 1998. Verocytotoxigenic *Escherichia coli* infection in dairy farm families. Can. Commun. Dis. Rep. 24–3.

World Health Organization (WHO). 1993. In: WHO guidelines for drinking water quality, 2nd ed. Vol. 1 – Recommendations. Geneva, WHO. pp. 8–29.

World Health Organization (WHO). 1999. Food safety – report by the Director-General. EB105/10, 2 December 1999. pp. 1–2.

World Health Organization (WHO). 2000. Bottled drinking-water [Online]. Fact Sheet No. 256, October, 2000. URL: http://www.who.int/inf-fs/en/fact256.html (Accessed: February 8, 2002).

Wood, J. D., G. A. Chalmers, R. A. Fenton, J. Pritchard, M. Schoonderwoerd, and W. L. Lichtenberger. 1991. Persistent shedding of *Salmonella enteritidis* from the udder of a cow. Can. Vet. J. 32:738–741.

Zhao, T., M. P. Doyle, P. Zhao, P. Blake, and F-M, Wu. 2001. Chlorine inactivation of *Escherichia coli* O157:H7 in water. J. Food Protect. 64:1607–1609.

Zottola, E. A., D. L. Schmeltz, and J. J. Jezeski. 1970. Isolation of salmonellae and other air-borne microorganisms in turkey processing plants. J. Milk Food Technol. 33:395–399.

CHAPTER 2

Pathogen Survival in Swine Manure Environments and Transmission of Human Enteric Illness—A Review[a]

SUMMARY

The influence of zoonotic pathogens in animal manure on human health and well-being as direct or indirect causes of human enteric illness is examined. Available international data are considered, but the study is focused on the developing situation in western Canada, where it is certain there will be further rapid growth in livestock numbers, particularly hogs. Major pathogens considered are *Escherichia coli* O157:H7, *Salmonella, Campylobacter, Yersinia, Cryptosporidium*, and *Giardia.* Canada is now the leading exporter of pork internationally, but recent increases in production contrasts with constant domestic levels of pork consumption and declining levels of foodborne illness caused by pork. Effects of increased levels of manure production are not quantifiable in terms of effects on human health. The presence of major pathogens in manure and movement to human food sources and water are considered on the basis of available data. Survival of the organisms in soil, manure, and water indicate significant variability in resistance to environmental challenge which are characteristic of the organisms themselves. Generally, pathogens survived longer in environmental samples at cool temperatures but differences were seen in liquid and solid manure. Based on actual data plus some data extrapolated from cattle manure environments, holding manure at 25°C for 3 months will render it free from the pathogens considered above.

[a] This chapter was published, by permission American Society of Agronomy, Crop Science Society of America and the Soil Science Society of America: Guan T. Y. and R. A. Holley (2003) Journal of Environmental Quality, 32:383–392.

This review evaluates the survival of zoonotic-based protozoan and bacterial pathogens in soil, water, and manure. We focus on several aspects related to the issue, including improper manure handling and foodborne illness, the effect of increased hog production, and environmental survival of the major zoonotic pathogens from swine and cattle.

MANURE HANDLING AND FOODBORNE ILLNESS

The danger of improper manure handling can be manifest as direct contamination of produce, water supplies, animals, or even humans. A summary of human enteric diseases where manure handling was implicated as the cause of infection is presented in Table 2.1. As shown in the table, manure application affects the safety of fresh produce and water supplies. Waterborne outbreaks and those associated with fresh produce have been on the rise in recent decades and will likely increase, in part due to increased surveillance. In most of these outbreaks the source of contamination was not confirmed. Although rarely identified as the cause, whenever manure was implicated in an outbreak, the results were serious (Table 2.1). In half of these outbreaks, mortality occurred. Since the source of contamination in most disease outbreaks is uncertain, the risks from improper manure handling are probably greatly underestimated.

In the field, produce can be directly contaminated via intentional application of raw manure as a fertilizer or indirectly contaminated from irrigation water which is accidentally contaminated with raw manure. These have serious consequences on ready-to-eat products such as unpasteurized apple cider and sprouts. Apples used for making cider can be contaminated when dropped on ground that is fertilized with manure. Pathogens on the skin of apples can be spread throughout the batch of cider (Besser et al. 1993). Sprout seeds may initially become contaminated on the farm through the use of manure as fertilizer, and subsequently during the sprouting process, pathogens are increased to high levels within the seed lot (Taormina et al. 1999). Despite the recognition of this hazard, there has been an increase in outbreaks associated with these products in recent years. Root crops such as radishes and carrots as well as leafy vegetables like lettuce, where the edible part touches the soil, also present a great risk for potential health problems. Two outbreaks of *Escherichia coli* O157:H7 associated with lettuce were traced back to organic growers who probably contaminated the produce with cow manure containing *E. coli* (Nelson 1997). One outbreak of *Citrobacter freundii* infections associated with parsley originated from an organic garden in which pig manure was used (Tschape et al. 1995). In the American organic standards and guidelines, composted manure is recommended for use by organic growers. Raw manure may also be used but not within 90–120 days prior to harvest, depending on the type of crop (Riddle et al. 1999). Foodborne illness surveillance data from 1990 to 1998 in the U.S. showed that contaminated produce (fruit and vegetables including juices and salads) account for about 24% of illness outbreaks (41% of cases), meats account

Table 2.1. Examples where manure has been implicated as the source of pathogens.

Location and date	Type of manure	Pathogen(s)	Vehicle(s)	Human morbidity and mortality	Circumstances leading to water/food contamination	Reference
1979–1981, Maritime Provinces, Canada	Sheep manure[a]	*Listeria monocytogenes*	Cabbage	34 cases of perinatal listeriosis and 7 cases of adult disease[b]	Cabbage was grown in fields fertilized with both composted and raw manure from a flock of sheep of which 2 had died of listeriosis, one in 1979 and one in 1981	Schlech et al. (1983)
July 1985, UK	Cow manure[a]	*Escherichia coli* O157:H7	Handling of potatoes	49 cases including 1 death	One load of potatoes became contaminated with cow manure before distribution	Morgan et al. (1988)
Oct 24–Nov 20, 1991, southeastern Massachusetts, USA	Cattle manure[a]	*E. coli* O157:H7	Unpasteurized, unpreserved, fresh-pressed apple cider	23 cases and no deaths	Apple cider was made from dropped apples collected from the ground which were probably contaminated with cattle manure	Besser et al. (1993)
Sept 23–Oct 1, 1992 Maine, USA	Cow & calf manure[a]	*E. coli* O157:H7	Vegetables	1 death and 4 cases	Vegetables were grown in 1st patient's garden which was fertilized all summer with manure from a cow and calf	Cieslak et al. (1993)
October 1992, Africa	Cattle carcass and manure[a]	*E. coli* O157:H7	Drinking water	Thousands of cases and some deaths	Surface waters probably contaminated by cattle carcasses and dung having been washed into rivers and dams by heavy rains	Isaacson et al. (1993)

(*Continued*)

Table 2.1. (*Continued*)

Location and date	Type of manure	Pathogen(s)	Vehicle(s)	Human morbidity and mortality	Circumstances leading to water/food contamination	Reference
March–April 1993, Milwaukee, USA	Cattle manure[a]	*Cryptosporidium*	Municipal water	403,000 cases	Rivers swelled by spring rains and snow runoff transported oocysts from cattle along the rivers into Lake Michigan and then to the treatment plant intake	Mac Kenzie et al. (1994)
October 1993, Maine, USA	Calf manure	*Cryptosporidium*	Fresh-pressed apple cider	160 primary cases	Apples contaminated by calf feces on the ground	Millard et al. (1994)
Summer, early 90s, Germany	Hog manure[a]	*Citrobacter freundii*	Sandwich prepared with green butter made with contaminated parsley	1 death, 8 HUS, 8 gastroenteritis cases and 20 asymptomatic cases	Parsley grown in a private organic garden in which pig manure was used	Tschape et al. (1995)
4 June 1995, ON, Canada	Cattle manure[a]	*E. coli* O157:H7	Well water from a shallow dug well on an dairy farm	1 case of bloody diarrhea	Design and location of a well allowed manure-contaminated water to flow into the well	Jackson et al. (1998)
June 1996, New York, USA	Poultry manure[a]	*Salmonella* Hartford & *Plesiomonas shigelloides*	Food prepared with contaminated water	About 30 cases and 1 hospitalization	An unprotected shallow dug well may have received surface runoff from surrounding tilled, manured farmland following rainfall	CDC (1998)

(*Continued*)

Location/Date	Source of contamination	Pathogen	Water/Source	Cases/Deaths	Description	Reference
June–July 1997, Somerset, UK	Cow manure[c]	*E. coli* O157	Contaminated mud at an open-air music festival	8 cases	Infected cattle (650 cows) grazed on the site 2 days prior to the festival	Crampin et al. (1999)
Summer 1999, Scotland, UK	Sheep manure[c]	*E. coli* O157	Untreated drinking water	6 cases	Contamination of untreated, unprotected private water source in a rural area where sheep and deer grazed freely	Licence et al. (2001)
May–June 2000, ON, Canada	Cattle manure[c]	*E. coli* O157:H7 & *Campylobacter* spp.	Treated municipal water	6 deaths and 1,346 reported cases	Pathogens from cattle manure on adjacent farms entered municipal well following heavy rains and flooding	Health Canada (2000)
March–May 2001, SK, Canada	Animal or human waste[a]	*Cryptosporidium parvum*	Municipal drinking water	1,907 cases and no deaths	Surface river water was probably contaminated from some point upstream	Health Canada (2001)

[a]Suspected as the source of contamination.
[b]9 fetal deaths, 7 infant deaths, and 2 adult deaths.
[c]Confirmed as the source of contamination.

for about 29% (20% of cases) and seafood about 14% (8% of cases) (Griffiths 2000). These data demonstrated that the outbreaks involving produce resulted in a greater number of reported cases than outbreaks involving meats.

Proper composting of manure can yield safe fertilizer. In Canada, regulations for treating animal manure are almost non-existent. Farmers are provided with guidelines for storing and spreading manure, however, the guidelines are voluntary. Canadian commercial compost standards require that during windrow composting a temperature of 55°C or greater is maintained for at least 15 days during the composting period, and that during the period the compost is turned at least five times (CCC 2002). For industrial composting systems where the process is closely monitored and controlled, consistent elimination of pathogens can be achieved. However, in many farm composting systems there is less control over the process and it is more difficult to ensure uniform exposure to high temperature, which may result in the survival of some pathogens (Patriquin 2000). On the farm, there is also a likelihood of reintroduction of pathogens by sequential addition of new manure during the composting process. Allowing time for proper composting is critical. Two to four month composting times have been suggested for backyard composts to get rid of *E. coli* O157:H7 (ENN 1997). Composting of manure is obviously important for its use on food crops, but may also be important for forage crops in order to reduce levels of *E. coli* O157:H7 in livestock. *E. coli* O157:H7 is resident much longer in manure than in the live animals, and thus manure-contaminated materials are thought to be a source for reinfection of livestock with *E. coli* O157:H7 (Kudva et al. 1998).

Land application of raw manure also results in contamination of agricultural run-off and water supplies. Contaminated drinking water has the potential to cause extensive outbreaks due to the large populations served by many distribution systems. The Walkerton (ON) and Milwaukee (WI) outbreaks are two well known examples. In Walkerton it was confirmed that human pathogens from cattle manure on adjacent farms entered the municipal water supply following heavy rains and flooding (Health Canada 2000). The outbreak resulted in six deaths. In the Milwaukee outbreak, it was estimated that more than 100 deaths occurred over the two years following the massive outbreak there. Many of these were due to chronic complications especially in immunocompromised persons (Hoxie et al. 1997).

The extent of water contamination in Canada due to agricultural practices has not been well studied. In one report it was found that bacterial contamination of surface water occurred at a single field site in Ontario due to liquid manure spread using accepted practices over a two year period (Joy et al. 1998). Results showed that significant numbers of bacteria could reach the surface water by infiltrating through the soil and traveling through sub-surface tile drains to the receiving water. Rainfall shortly after manure application was suggested to be the most important factor influencing bacterial contamination rather than spreading rate (volume applied per unit area) or condition of the field prior to spreading. Goss et al. (1998), examined well contamination problems resulting from the use of manure in Ontario, and reported that the number

Table 2.2. Hogs marketed in Canada by province, 1984–2000, ('000 head).[a]

Year	NF[b]	PEI	NS	NB	QC	ON	MB	SK	AB	BC	CANADA
1984	36.6	182.1	255.5	168.9	4,794.5	4,956.5	1,608.7	812.0	2,030.8	387.3	15,232.5
1988	28.6	194.5	243.1	141.7	4,739.0	4,982.2	2,124.2	1,054.1	2,397.8	401.5	16,306.8
1992	27.2	173.0	212.7	123.6	4,703.2	4,424.9	2,353.1	1,174.5	2,589.6	358.4	16,140.2
1996	7.2	190.3	218.5	134.1	5,463.4	4,586.2	3,149.1	1,195.1	2,681.8	332.7	17,958.4
1997	7.3	187.2	224.4	144.8	5,767.5	4,817.6	3,225.1	1,234.0	2,629.2	328.0	18,565.1
1998	6.1	198.7	204.8	173.7	6,262.8	6,436.5	3,967.6	1,437.4	2,804.0	357.7	21,045.2
1999	6.8	208.1	233.4	181.9	6,778.1	5,844.4	4,978.7	1,410.1	3,105.6	342.1	23,089.2
2000	7.0	201.7	227.4	202.8	6,869.4	6,068.5	5,337.1	1,496.9	3,326.2	319.1	24,056.1

[a]Statistics Canada cited by CPC (2001).
[b]Provinces: NF-Newfoundland, PEI-Prince Edward Island, NS-Nova Scotia, NB-New Brunswick, QC-Quebec, ON-Ontario, MB-Manitoba, SK-Saskatchewan, AB-Alberta, BC-British Columbia.

of wells showing bacterial contamination increased from 15% to 25% between 1955 and 1992. In Manitoba, early in the summer of 2000, the community of Balmoral in the regional municipality (RM) of Rockwood was advised to boil its drinking water. It was found that of the 75 wells serving the community, 86% contained coliform and/or *E. coli* bacteria (CMIP 2001a). In September 2000, another boil water advisory was issued for the community of Haywood in the RM of Grey in the province after the finding that 90% of 55 wells sampled were contaminated by bacteria (CMIP 2001b). However, the sources of contamination in both incidents were not determined.

In Manitoba, ≥23.6 million tonnes of manure are produced annually, according to 1997 statistics (MRWQ 1999). The largest single manure source in the province was cattle grazed on rangeland where manure is not collected. When manure production in Manitoba is broken down by livestock group, cattle and calves were responsible for 84.5%, hogs 14% and others 1.5% (1997 data). Manure production will increase as the hog industry in the province continues to grow (MRWQ 1999). Manitoba is the 3rd largest hog producer in the country after Quebec and Ontario (Table 2.2). Most manure-associated outbreaks have implicated bovine manure more frequently than other types (Table 2.1). This has led to more study of pathogens in bovine than in hog manure. The amount of information on pathogens in swine manure is very limited. A study on Ontario livestock farms in 1996 (Fleming 1999) found that *Cryptosporidium* appeared to be more prevalent in swine than in dairy manure. This study showed that 26% of all swine liquid manure samples tested positive for the protozoan, compared to 8.1% for dairy solid manure and 7.3% for dairy liquid manure. For each of the three farm types (swine farms with liquid manure, dairy farms with solid manure, and dairy farms with liquid manure), 50–55% of the farms tested positive for the protozoan at least once in the fresh manure (from young pigs or calves). In contrast, 75% of the swine farms with liquid manure storages tested positive at least once, compared to 20% of dairy farms with solid manure storages and none of the liquid dairy manure storages. The author's later study in 1998 on Ontario swine farms suggested that oocysts were present in about

50% of manure samples, and of those that tested positive, at least some viable organisms were present 60% of the time (Fleming 1999).

EFFECT OF INCREASED HOG PRODUCTION

Incidents (outbreaks) and cases of pork-associated human enteric illness in Canada, from 1975 to 1995, are presented in Figs 2.1 and 2.2. There appears to be a trend toward a decrease in terms of both total incidents and cases. Per capita disappearance (an estimate of pork consumption) in Canada, measured by retail weight sold, did not vary much between 1982 and 2000 (high of 22.26 kg in 1983; low of 19.27 kg in 1997), despite an increase in national hog production over the same period of time (Figs 2.3 & 2.4). There appears to be no relationship between human enteric illness from pork and total hog production in Canada. Pork is the most popular meat in the world, comprising 43% of the world's meat consumption in 1997 (Binnie 1999). Canada is the largest pork exporter in the world, contributing 23% of world exports in year 2000 (USDA 2001a). In 2000, Canada exported 40% of its hog production compared to 21% in 1982 (Statistics Canada 2000, 2001). It appears that the increase in national hog production has been used to increase hog exports while domestic pork consumption has remained stable. Thus the more than doubling of hog production in Manitoba since 1992 has not affected rates of human illness from pork consumption (Figs 2.2 & 2.4). Among other things, the data indicate that the safety of pork (as a vehicle for foodborne illness) has either remained unchanged or improved during this period.

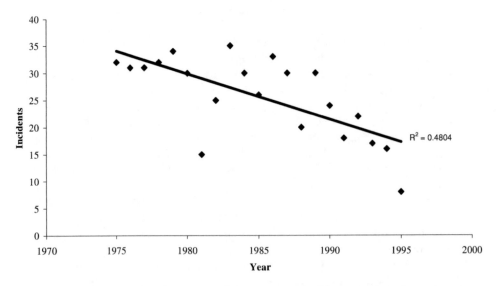

Figure 2.1. Incidents of pork-associated human enteric illness in Canada, 1975–1995. [Tood (1990, 1991), Todd & Harboway (1994), Todd & Chatman (1996, 1997, 1998) and Todd et al. (2000).]

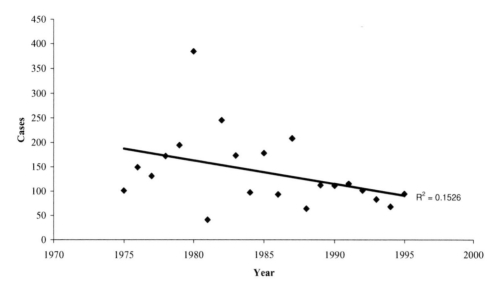

Figure 2.2. Cases of pork-associated human enteric illness in Canada, 1975–1995. [Tood (1990, 1991), Todd & Harboway (1994) Todd & Chatman (1996, 1997, 1998) and Todd et al. (2000).]

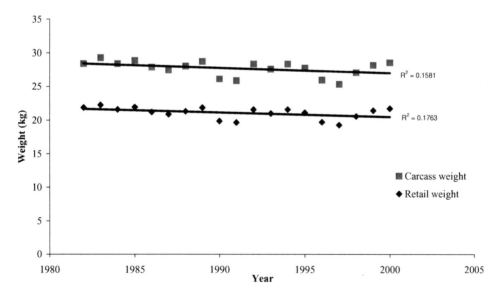

Figure 2.3. Per capita disappearance of pork in Canada, 1982–2000. [Statistics Canada (2000, 2001). Does not estimate trim at wholesale and retail or table waste.]

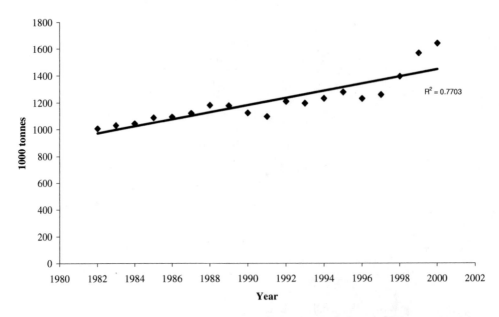

Figure 2.4. Canadian hog production, 1982–2000. [Statistics Canada (2000, 2001).]

In the United States, an estimated 14.6% of all known causes of foodborne illness outbreaks are related to pork consumption (USDA 2001b). The U.S. hog carcass baseline data for *Salmonella* (8.7%), *Campylobacter* (31.5%), and *Listeria monocytogenes* (7.4%) suggest the widespread distribution of human pathogens in U.S. pork (FSIS 1996). *Salmonella* levels were selected as the performance standard for the national HACCP program for slaughter plants and establishments that produce raw ground products. In 2001, the *Salmonella* prevalence in swine was reduced to 1.6% in large plants (those with 500 or more employees), 3.5% in small plants (those with 10 or more but fewer than 500 employees), and 4.4% in very small plants (those with fewer than 10 employees) (FSIS 2002). The U.S. statistics from 1973 to 1992 showed that there has been a drop in the total number of foodborne disease outbreaks attributed to pork (Table 2.3) (CDC cited by USDA 1997a). However the total number of cases in these outbreaks was not available. Foodborne illness outbreaks and cases from USDA-regulated pork products did not show any apparent trend from 1990 to 1998 (CSPI 2000). Similar to Canadian data, per capita pork consumption in the U.S. remained fairly stable from 1992 to 2000 despite a gradual increase in pork production (Table 2.4). No obvious relationship between pork consumption patterns, pork production, and pork-related human enteric illness can be established.

It was reported that there are up to 125 infectious agents found in hogs world-wide (D'Allaire et al. 1999). The list includes the well-known *E. coli* O157:H7. To date, however, there has been no direct evidence of human *E. coli* O157:H7 infections originating from pigs in North America. This is in agreement

**Table 2.3. Foodborne outbreaks of
human illness from pork in the
United States, 1973–1992[a]**

Period	Total number of outbreaks
1973–1977	119
1978–1982	86
1983–1987	47
1988–1992	29

[a]CDC cited by USDA (1997a).

with the zero prevalence of the pathogen in the U.S. baseline data for swine (FSIS
1996). Until recently, toxigenic *E. coli* O157:H7 was not thought to be present in
North American pigs. Gyles et al. (2002) reported for the first time, isolation of
E. coli O157:H7 from growing pigs (animal prevalence of 3.03% and herd prevalence of 6.82%) in Ontario, Canada. Earlier on, isolation of *E. coli* O157:H7 from
pigs had been reported in Chile (Gyles et al. 2002), Japan (Nakazawa et al. 1999),
the Netherlands (Heuvelink et al. 1999), and Norway (Johnsen et al. 2001). The
prevalence of the pathogen in pigs varied in the latter countries, ranging from
0.1–1.4%. In a survey of pigs in England, non-toxigenic *E. coli* O157 was isolated
from 0.4% of fecal samples collected from 1000 pigs after slaughter (Chapman
et al. 1997). An earlier Canadian study (Read et al. 1990) reported the prevalence
of non-O157:H7 verocytotoxigenic *E. coli* (VTEC) in 10.6% of pork samples
collected from meat processing plants in Ontario, and a number of the serotypes
of the isolates have been associated with human disease. In Denmark, VTEC
were also isolated from 7.5% of pigs (Beutin et al. 1993). A recent study by
DesRosiers et al. (2001) found that certain non-O157:H7 porcine VTEC such
as O91:NM, a serotype that has been associated with hemorrhagic colitis in
humans, may potentially infect humans. Therefore, VTEC other than *E. coli*

**Table 2.4. Per capita pork consumption[a] and pork production[b]
in the U.S., 1992–2000.**

Year	Per capita pork consumption—carcass weight (kg)	Pork Production—carcass weight (1000 tonnes)
1992	31.0	7,817
1993	30.6	7,751
1994	31.0	8,027
1995	30.7	8,097
1996	29.0	7,764
1997	28.7	7,835
1998	31.0	8,623
1999	31.8	8,758
2000	31.0	8,596

[a]USDA (1997b, 1999, 2001a).
[b]USDA (1997c, 2000, 2001a, 2002).

O157:H7 may be the more important human pathogens in pigs. As in the U.S., *Salmonella* in pork is also under surveillance in Denmark, Sweden, Norway, and Finland. In Denmark, *Salmonella enterica* serovar Typhimurium was isolated more frequently from pork than from broilers (Hald & Wegener 1999). The majority (70%) of the human *S.* Typhimurium infections were believed to be attributed to pork while only 10% was acquired from poultry. In addition to VTEC and *Salmonella*, hog manure is also known to harbour pathogenic *Yersinia enterocolitica*. Pork is considered to be the only source of human yersiniosis in Denmark as the majority of Danish pig herds harbour *Y. enterocolitica* (herd prevalence of 90% and within-herd prevalence of 80%) (Anon 2001).

ENVIRONMENTAL SURVIVAL OF THE MAJOR ZOONOTIC PATHOGENS FROM SWINE AND CATTLE

The survival of several major zoonotic pathogens under different environmental conditions is summarized in Table 2.5. Predictions are made where data do not exist. In general, zoonotic pathogens appear to survive longer in water, followed by soil and manure. In each of these environments, they survive better at lower than at higher temperatures.

Survival in Water

Survival of *Escherichia coli* O157:H7 in water was studied by Wang & Doyle (1998). Greatest survival was reported in filtered, autoclaved municipal water and least in lake water. Regardless of the water source, survival was greatest at 8°C and least at 25°C. The pathogen survived for at least 91 days at 8°C, but was not detectable within 49 to 84 days at 25°C. It was also demonstrated that *E. coli* O157:H7 can enter a viable but nonculturable (VBNC) state in water (Wang & Doyle 1998). This VBNC state of the pathogen in the natural environment poses an epidemiological concern since the source of contamination can be overlooked during outbreak investigations where conventional recovery methods are used. Nonetheless, the concept of bacteria adopting a VBNC state as a survival strategy in adverse environmental conditions has not gained universal acceptance (Weichart 1999).

The fate of *Salmonella* species (including *Salmonella enterica* serovar Typhimurium DT 104) in river water was examined by Santo Domingo et al. (2000). The authors inoculated river water and steam sterilized (autoclaved) river water with 8 Log_{10} CFU/ml *Salmonella*. After 45 days at room temperature, they observed that the number of *Salmonella* was reduced by more than 3 logs in the untreated water while the number of these organisms was reduced by two to three logs in the treated water. However, when the VBNC cells were examined, less than one log of reduction was observed after 45 days in both types of water. Lengthy survival of *Salmonella* spp. in natural water was also recorded in Mitscherlich & Marth (1984) where a survival time of at least 152 days was

Table 2.5. Survival of enteric pathogens in natural environments.

Environment	Temperature	E. coli O157:H7[a]	Salmonella[b]	Yersinia enterocolitica[c]	Campylobacter[d]	Giardia[e]	Cryptosporidium[e]
Natural water	Frozen (−4°C)	[>91 d][f,g]	[152 d]	[448 d]	[120 d]	<7 d	>84 d
	Cold (4–8°C)	>91 d	[152 d]	448 d	8–120 d	77 d	>84 d
	Warm (20–30°C)	49–84 d	45–152 d	10 d	<2 d	14 d	70 d
Soil	Frozen (−4°C)	[99 d]	[63 d]	[10 d]	[20 d]	<7 d	>84 d
	Cold (4–6°C)	99 d	63 d	[10 d]	20 d	49 d	56 d
	Warm (20–30°C)	56 d	>45 d	10 d	10 d	7 d	28 d
Cattle manure	Frozen (−20 or −4°C)	>100 d	[48 d]	[10 d]	[21 d]	<7 d	>84 d
	Cold (4–5°C)	70 d	[48 d]	[10 d]	12–21 d (human feces)	7 d	56 d
	Warm (20–37°C)	49–56 d	48 d	[10 d]	3 d	7 d	28 d
Cattle manure slurry/liquid	4, 20, or 37°C	27–60 d	19–60 d	[10 d]	3 d	[7 d]	[28 d]

[a]Reference(s) for water: Wang & Doyle (1998); soil: Bolton et al. (1999), Mubiru et al. (2000); cattle manure: Kudva et al. (1998), Wang et al. (1996); slurry: Himathongkham et al. (1999).

[b]Reference(s) for water: Santo Domingo et al. (2000), Mitscherlich & Marth (1984); soil: Zibilske & Weaver (1978); cattle manure & slurry: Himathongkham et al. (1999).

[c]Reference(s) for water: Karapinar & Gonul (1991), Chao et al. (1988); soil: Chao et al. (1988).

[d]Reference(s) for water: Buswell et al. (1998), Rollins & Colwell (1986); soil: Mitscherlich & Marth (1984); cattle manure & slurry: Valdes-Dapena Vivanco & Adam (1983), Blaser et al. (1980), Mitscherlich & Marth (1984).

[e]Reference for water, soil and cattle manure: Olson et al. (1999).

[f]d—day(s).

[g]Value in [ ] is value predicted based on existing values from other environments.

noted in sterilized well water at 18–20°C. It is clear that *Salmonella* species can survive for a long period of time in natural water bodies, and possibly even longer at lower temperatures.

Yersinia enterocolitica is a psychrotrophic organism, which underlines its significance in refrigerated foods where it can grow. Its lengthy survival in water was noted by Karapinar & Gonul (1991) who detected survivors in sterile spring water after 64 weeks at 4°C. Survival was greatly reduced with increasing temperatures. For example, Chao et al. (1988) reported viable numbers after 6 days at 16°C in river water, and 10 days at 30°C in groundwater. The difference was explained by the presence of eukaryotic predators and toxic materials (e.g. biological toxins, chemical toxicants), which are usually higher in surface water. Terzieva & McFeters (1991) also noted extended survival at reduced temperatures, with the pathogen surviving at least 14 days at 16°C.

In water, *Campylobacter* survived for 8 to 28 days at 4°C (Buswell et al. 1998; Terzieva & McFeters 1991; Blaser et al. 1980). Survival times were similar at 16°C (Terzieva & McFeters 1991) but were greatly shortened as the temperature increased to 22°C and above. At 22 and 37°C, the organism only survived for 43 and 22 hours, respectively (Buswell et al. 1998). The VBNC state of *Campylobacter jejuni* in water was demonstrated by Rollins & Colwell (1986). They found that stream water held at low temperature (4°C) sustained significant numbers of VBNC *C. jejuni* for more than 4 months.

Giardia cysts are much more susceptible to environmental stress than *Cryptosporidium* oocysts. A study by Olson et al. (1999) showed that temperatures as low as −4°C inactivated *Giardia* cysts in water while *Cryptosporidium* oocysts remained viable for >12 weeks. At 4°C in water *Giardia* cysts were infective for 11 weeks while *Cryptosporidium* oocysts again survived for >12 weeks. At 25°C *Giardia* cysts were noninfective in water within 2 weeks but *Cryptosporidium* oocysts were infective up to 10 weeks.

In cold water (4–8°C), *Y. enterocolitica* has the greatest survival among all pathogens whereas *Campylobacter* has possibly the least. In warm water (20–30°C), *Salmonella* survives best, followed by *E. coli* O157:H7, *Cryptosporidium*, *Giardia*, *Y. enterocolitica*, and lastly *Campylobacter*.

Survival in Soil

E. coli O157:H7 can survive in soil for a long period of time depending on the soil type. In the laboratory, the organism survived for at least 8 weeks in moist soil at 25°C (Mubiru et al. 2000). Under fluctuating environmental temperatures (−6.5 to 19.6°C), the organism can be detected for up to 99 days (Bolton et al. 1999). Among the physical and chemical properties of soil, soil moisture is a major factor determining bacterial survival. Greater survival is often associated with moist soils, thus rainfall is a factor which favours bacterial survival.

Salmonella is relatively persistent in soil compared to other pathogens. When inoculated at 8 Log_{10} CFU/g into moist soil which was then stored at 20°C, less than 2 log reductions were observed after 45 days (Guo et al. 2002). These findings are consistent with the results of an early study by Zibilske & Weaver (1978) who reported survival of *S.* Typhimurium in soil for 42 days at 22°C, and for 63 days at 5°C. Under natural environmental conditions, *S.* Typhimurium was isolated up to 14 days from agricultural soil amended with *Salmonella*-contaminated hog manure slurry which was spread in spring (Baloda et al. 2001). The authors also cited their unpublished data which showed that under controlled conditions in terrestrial ecosystems, *S.* Typhimurium DT 104 and DT 12 could survive up to 299 days, although the temperature was not noted.

In the laboratory, *Y. enterocolitica* survived for 7 days in soil at 30°C (Chao et al. 1988). Survival times could be prolonged to 10 days in pH-adjusted soil (final pH 6.5) and air drying of the pH-adjusted soil reduced the number of survivors. In an early study, *Campylobacter intestinalis* survived in non-sterilized soil for 20 and 10 days at 6° and 20–37°C, respectively (Lindenstruth et al. 1949 cited by Mitscherlich & Marth 1984).

While *Giardia* is sensitive to freezing of soil, *Cryptosporidium* is resistant. Olson et al. (1999) reported *Giardia* cysts in soil were non-infective after seven days at −4°C, but *Cryptosporidium* could survive for >12 weeks. Infective *Giardia* cysts and *Cryptosporidium* oocysts were recovered from soils maintained at 4°C for up to 8 weeks. At 25°C, *Giardia* cysts were inactivated at one week in soil whereas *Cryptosporidium* oocysts survived and were infective for 4 weeks. *Cryptosporidium* oocysts were degraded more rapidly in soil containing natural microorganisms than in sterile soil (Olson et al. 1999). It has been suggested that soil sterilization destroys competing indigenous microflora, increases available nutrients for microorganisms and reduces concentrations of inhibitory compounds unstable at treatment temperatures (Mitscherlich & Marth 1984).

In cold (4–6°C) soil, most pathogens can survive for at least a month. Among all pathogens considered, the greatest survival was found for *Cryptosporidium* in frozen soil and *E. coli* O157:H7 and *Salmonella* in warm (20–30°C) soil (Table 2.5).

Survival in Manure

When exposed to fluctuating environmental conditions, *E. coli* O157:H7 could survive for more than 1 year in non-aerated ovine manure (Kudva et al. 1998). In aerated ovine and bovine manure piles held in the natural environment, the organism survived for 4 months and 47 days, respectively. Aeration was believed responsible for accelerating the drying of manure, and this probably resulted in the reduction of the number of pathogens (Kudva et al. 1998). Under controlled laboratory conditions, the pathogen survived for at least 100 days in bovine manure frozen at −20°C and 100 days in ovine manure incubated at 4 or 10°C. These results were in agreement with those of Wang et al. (1996) who also reported longer survival in manure (bovine) at lower temperatures. Their studies showed *E. coli* O157:H7 survived for 70 days at 5°C, 56 days at 22°C, and 49 days at 37°C.

**Table 2.6. Decimal reduction time (DRT) values (days) of *E. coli* O157:H7 and
S. Typhimurium in cattle manure and manure slurry.[a]**

Material	Temperature	*E. coli* O157:H7	*S.* Typhimurium
Manure	4°C	9.0/18.6[b]	12.7/20.3
	20°C	21.6/13.5	24.7/9.4
	37°C	8.9/3.6	8.4/1.7
Fresh manure slurry	4°C	21.5	16.4
	20°C	14.8	12.7
	37°C	3.2	2.4
Slurry from old manure	4°C	38.8	65.8
	20°C	7.7	2.8
	37°C	2.2	2.5

[a]Modified from Himathongkham et al. (1999).
[b]Top layer/middle and bottom layers (total depth of 8 cm) of manure.

Survival data for *Salmonella* spp. in various animal manures are reported in
Mitscherlich & Marth (1984). In pig slurry, *S.* Senftenberg survived for 14 days
at 8°C, 8 days at 20°C, and <8 days at 37°C (Muller 1973 cited by Mitscherlich
& Marth 1984). In cattle slurry, *S.* Typhimurium was not detectable after 19 days
at 37°C, but survived for at least 60 days at 4 and 20°C (Himathongkham et al.
1999). In cattle manure, the organism survived for 48 days at 37°C. The authors observed an exponential linear destruction of *S.* Typhimurium and *E. coli*
O157:H7 in cattle manure and manure slurry stored at 4, 20, or 37°C. The decimal reduction times (DRT; times required for 90% reduction) of the pathogens
are presented in Table 2.6. DRT values may be useful in risk assessments to
predict how long manure should be held before application to fields. It appears
that both pathogens survived less well in liquid manure than in solid manure at
20 and 37°C while the opposite was true at 4°C. Therefore, handling manure as
a liquid may be a better alternative as less time is required to kill the pathogens.
Further, greater survival occurred at 20 and 37°C in fresh manure slurry compared to old manure slurry (slurry made from manure which was stored for
months and had become dried). This indicates that manure stored at warm temperatures for a period of time is an unfavorable environment for survival of
the pathogens. This is probably because stressed microorganisms survive for
shorter periods of time at warmer than cooler temperatures where metabolic
rates are slowed (Montville & Matthews 2001). In addition, these findings stress
the importance of spreading manure to fields only during warmer months.

There has been little previously reported data concerning survival of
Campylobacter and *Y. enterocolitica* in livestock manure. An early study (Zintz
1955 cited by Mitscherlich & Marth 1984) reported that *C. jejuni* survived for
3 days in both cattle feces and liquid cattle manure, and for 2 days in liquid
swine manure at room temperature. At 4°C in human feces, *C. jejuni* could be
recovered from 12 to 21 days (Valdes-Dapena Vivanco & Adam 1983; Blaser et al.
1980). A study by Kearney et al. (1993) examined the efficiency of a full scale,
continuous (daily fed) anaerobic digester on the survival of pathogenic bacteria. They found that *C. jejuni* was the most resistant bacterium to the anaerobic

digestion of cattle slurry (supplemented with pig, hen and potato waste) at 28°C, followed by *S*. Typhimurium, and *Y. enterocolitica*.

As in water and soil, *Giardia* survives less well than *Cryptosporidium* in manure. *Giardia* cysts in cattle feces were noninfective within a week following freezing at −4°C and were infective for only one week at 4 and 25°C (Olson et al. 1999). On the other hand, *Cryptosporidium* oocysts remained infective for >12 weeks at −4°C, 8 weeks at 4°C, and 4 weeks at 25°C.

CONCLUSIONS

Of those pathogens considered, *E. coli* O157:H7 was the most persistent organism in cattle manure regardless of the temperature and manure form (solid or slurry). On the contrary, *Campylobacter* and *Giardia* were the weakest survivors in manure. Readers can refer to Wang & Doyle (1998) and Kudva et al. (1998) for details on environmental survival of *E. coli* O157:H7. It is clear that storing manure as a slurry, solid, or compost before it is distributed on fields results in a significant reduction in pathogen concentration. As most pathogens survive freezing or low temperatures for significant periods of time, untreated manure should not be distributed on fields where there is a potential for runoff. The lack of scientific information on the survival of human pathogens in swine manure is an impediment in developing directions for changes to improve its handling. There is a pressing need to close information gaps on this subject. We hypothesize that it should be possible to eliminate the major bacterial and protozoan pathogens from bulk liquid manure holding systems by storage at 25°C for 3 months. This recommendation is primarily based on cattle manure data and may change in other animal manure systems. While this approach may be inconvenient in some climates it should, nonetheless, serve as a guide for future work to evaluate additional factors (e.g. pH) which could be used to optimize the lethal effects of the time-temperature relationship upon pathogens in stored manures. We contend that elimination of pathogens from manures used as fertilizer is a critical control point for managing the pathogen problem on crops used as feed and food and in managing the microbial safety of the water supply.

REFERENCES

Anonymous. 2001. Annual report on zoonoses in Denmark 2000. The Ministry of Food, Agriculture and Fisheries, Copenhagen, Denmark.

Baloda, S. B., L. Christensen, and S. Trajcevska. 2001. Persistence of a *Salmonella enterica* serovar Typhimurium DT12 clone in a piggery and in agricultural soil amended with *Salmonella*-contaminated slurry. Appl. Environ. Microbiol. 67:2859–2862.

Besser, R. E., S. M. Lett, J. T. Weber, M. P. Doyle, T. J. Barett, J. G. Wells, and P. M. Griffin. 1993. An outbreak of diarrhea and hemolytic uremic syndrome from *Escherichia coli* O157:H7 in fresh-pressed apple cider. J. Am. Med. Assoc. 269:2217–2220.

Beutin, L., D. Geier, H. Steinruck, S. Zimmermann, and F. Scheutz. 1993. Prevalence and some properties of verotoxin (shiga-like toxin)-producing *Escherichia coli* in seven different species of healthy domestic animals. J. Clin. Microbiol. 31:2483–2488.

Binnie, M. A. 1999. February designated Canadian pork month [Online]. Canadian Pork Council. URL: http://www.cpc-ccp.com/communic%5C99-02-01.pdf (Accessed: April 3, 2002).

Blaser, M. J., H. L. Hardesty, B. Powers, and W. L. Wang. 1980. Survival of *Campylobacter fetus* subsp. *jejuni* in biological milieus. J. Clin. Microbiol. 11:309–313.

Bolton, D. J., C. M. Byrne, J. J. Sheridan, D. A. McDowell, I. S. Blair, and T. Hegarty. 1999. The survival characteristics of a non-pathogenic strain of *Escherichia coli* O157:H7. In: G. Duffy, P. Garvey, J. Coia, Y. Wasteson, and D. A. McDowell (eds.). Verocytotoxigenic *E. coli* in Europe. 2. Survival and growth of verocytotoxigenic *E. coli*. Teagasc, The National Food Centre, Dublin, Ireland. pp. 28–36.

Buswell, C. M., Y. M. Herlihy, L. M. Lawrence, J. T. M. McGuiggan, P. D. Marsh, C. W. Keevil, and S. A. Leach. 1998. Extended survival and persistence of *Campylobacter* spp. in water and aquatic biofilms and their detection by immunofluorescent-antibody and—rRNA staining. Appl. Environ. Microbiol. 64:733–741.

Can.-MB Infrastr. Prog. (CMIP). 2001a. Infrastructure Program addressing Balmoral boil water advisory [Online]. URL: http://www.infrastructure.mb.ca/e/news010427t_b1.html (Accessed: April 5, 2002).

Can.-MB Infrastr. Prog. (CMIP). 2001b. Infrastructure Program addressing Haywood boil water advisory [Online]. URL: http://www.infrastructure.mb.ca/e/news010427t_b3.html (Accessed: April 5, 2002).

Canadian Pork Council (CPC). 2001. Statistics [Online]. URL: http://www.cpc-ccp.com/stats.html (Accessed: April 2, 2002).

Center for Science in the Public Interest (CSPI). 2000. Outbreaks traced to USDA-regulated foods, 1990–2000 [Online]. URL: http://www.cspinet.org/reports/outbreak_alert/appendix_b.htm (Accessed: April 1, 2002).

Centers for Disease Control and Prevention (CDC). 1998. *Plesiomonas shigelloides* and *Salmonella* serotype Hartford infections associated with contaminated water supply-Livingston County, New York, 1996. Morb. Mortal. Wkly. Rep. 47:394–396.

Chao, W-L, R-J Ding, and R-S Chen. 1988. Survival of *Yersinia enterocolitica* in the environment. Can. J. Microbiol. 34:753–756.

Chapman, P. A., C. A. Siddons, A. T. Cerdan Malo, and M. A. Harkin. 1997. A 1-year study of *Escherichia coli* O157:H7 in cattle, sheep, pigs, and poultry. Epidemiol. Infect. 119:245–250.

Cieslak, P. R., T. J. Barett, P. M. Griffin, K. F. Gensheimer, G. Beckett, J. Buffington, and M. G. Smith. 1993. *Escherichia coli* O157:H7 infection from a manured garden. Lancet, 342:367.

Composting Council of Canada (CCC). 2002. Setting the standard: a summary of compost standards in Canada [Online]. URL: http://www.compost.org/standard.html (Accessed: April 10, 2002).

Crampin, M., G. Willshaw, R. Hancock, T. Djuretic, C. Elstob, A. Rouse, T. Cheasty, and J. Stuart. 1999. Outbreak of *Escherichia coli* O157 infection associated with a music festival. Eur. J. Clin. Microbiol. Infect. Dis. 18:286–288.

D'Allaire, S., L. Goulet, and J. Brodeur. 1999. Literature review on the impacts of hog production on public health [Online]. Symposium of the Hog Environmental Management Strategy (Livestock Environmental Initiative). URL: http://res2.agr.ca/initiatives/manurenet/en/hems/hems2/h2_communications.html#Literature (Accessed: April 3, 2002).

DesRosiers, A., J. M. Fairbrother, R. P. Johnson, C. Desautels, A. Letellier, and S. Quessy. 2001. Phenotypic and genotypic characterization of *Escherichia coli* verotoxin-producing isolates from humans and pigs. J. Food Protect. 64:1904–1911.

Environmental News Network (ENN). 1997. Food gardeners urged to avoid fresh manure [Online]. URL: http://www.enn.com/news/enn-stories/1997/06/060397/06039711.asp (Accessed: April 4, 2002).

Fleming R. 1999. *Cryptosporidium*: is manure a contributor? [Online] Agri-food research in Ontario. URL: http://www.gov.on.ca/OMAFRA/english/research/magazine/summer99/pdfs/p22-23.pdf (Accessed: April 10, 2002).

Food Safety and Inspection Service (FSIS). 1996. Nationwide pork microbiological baseline data collection program: market hogs [Online]. URL: http://www.fsis.usda.gov/OPHS/baseline/markhog1.pdf and http://www.fsis.usda.gov/OPHS/baseline/markhog2.pdf (Accessed: April 11, 2002).

Food Safety and Inspection Service (FSIS). 2002. Progress report on *Salmonella* testing of raw meat and poultry products, 1998–2001 [Online]. URL: http://www.fsis.usda.gov/ophs/haccp/salm4year.htm (Accessed: May 1, 2002).

Goss, M. J., D. A. J. Barry, and D. L. Rudolph. 1998. Contamination in Ontario farmstead domestic wells and its association with agriculture: 1. Results from drinking water wells. J. Contam. Hydrol. 32:267–293.

Griffiths, M. W. 2000. The new face of foodborne illness. CMSA News, Can. Meat Sci. Assoc., Ottawa ON, March: 6–9.

Guo, X., J. Chen, R. E. Brackett, and L. R. Beuchat. 2002. Survival of *Salmonella* on tomatoes stored at high relative humidity, in soil, and on tomatoes in contact with soil. J. Food Protect. 65:274–279.

Gyles, C. L., R. Friendship, K. Ziebell, S. Johnson, I. Yong, and R. Amezcua. 2002. *Escherichia coli* O157:H7 in pigs. Expert Committee on Meat and Poultry Products, Canada Committee on Food, Can. Agri-Food Res. Council, September 16–17, Guelph ON.

Hald, T., and H. C. Wegener. 1999. Quantitative assessment of the sources of human salmonellosis attributable to pork [Online]. Proceedings, 3rd international symposium on the epidemiology and control of *Salmonella* in pork, Washington D. C., August 5–7, 1999. URL: http://www.isecsp99.org (Accessed: April 17, 2002).

Health Canada. 2000. Waterborne outbreak of gastroenteritis associated with a contaminated municipal water supply, Walkerton, Ontario, May–June 2000. Can. Comm. Dis. Rep. 26:170–173.

Health Canada. 2001. Waterborne cryptosporidiosis outbreak, North Battleford, Saskatchewan, spring 2001. Can. Comm. Dis. Rep. 27:185–192.

Heuvelink, A. E., J. T. M. Zwartkruis-Nahuis, F. L. A. M. van den Biggelaar, W. J. van Leeuwen, and E. de Boer. 1999. Isolation and characterization of verocytotoxin-producing *Escherichia coli* O157 from slaughter pigs and poultry. Int. J. Food Microbiol. 52: 67–75.

Himathongkham, S., S. Bahari, H. Riemann, and D. Cliver. 1999. Survival of *Escherichia coli* O157:H7 and *Salmonella typhimurium* in cow manure and cow manure slurry. FEMS Microbiol. Lett. 178:251–257.

Hoxie, N. J., J. P. Davis, J. M. Vergeront, R. D. Nashold, and K. A. Blair. 1997. Cryptosporidiosis-associated mortality following a massive waterborne outbreak in Milwaukee, Wisconsin. Am. J. Public Health, 87:2032–2035.

Isaacson, M., P. H. Canter, P. Effler, L. Arntzen, P. Bomans, and R. Heenan. 1993. Haemorrhagic colitis epidemic in Africa. Lancet, 341:961.

Jackson, S. G., R. B. Goodbrand, R. P. Johnson, V. G. Odorico, D. Alves, K. Rahn, J. B. Wilson, M. K. Welch, and R. Khakhria. 1998. *Escherichia coli* O157:H7 diarrhoea associated with well water and infected cattle on an Ontario farm. Epidemiol. Infect. 120:17–20.

Johnsen, G., Y. Wasteson, E. Heir, O. I. Berget, and H. Herikstad. 2001. *Escherichia coli* O157:H7 in faeces from cattle, sheep and pigs in the southwest part of Norway during 1998 and 1999. Int. J. Food Microbiol. 65:193–200.

Joy, D. M., H. Lee, C. M. Reaume, H. R. Whiteley, and S. Zelin. 1998. Microbial contamination of subsurface tile drainage water from field applications of liquid manure. Can. Agric. Engin. 40:153–160.

Karapinar, M., and S. A. Gonul. 1991. Survival of *Yersinia enterocolitica* and *Escherichia coli* in spring water. Int. J. Food Microbiol. 13:315–320.

Kearney, T. E., M. J. Larkin, J. P. Frost, and P. N. Levett. 1993. Survival of pathogenic bacteria during mesophilic anaerobic digestion of animal waste. J. Appl. Bacteriol. 75: 215–219.

Kudva, I. T., K. Blanch, and C. J. Hovde. 1998. Analysis of *Escherichia coli* O157:H7 survival in ovine or bovine manure and manure slurry. Appl. Environ. Microbiol. 64:3166–3174.

Licence, K., K. R. Oates, B. A. Synge, and T. M. S. Reid. 2001. An outbreak of *E. coli* O157 infection with evidence of spread from animals to man through contamination of a private water supply. Epidemiol. Infect. 126:135–138.

Mac Kenzie, W. R., N. J. Hoxie, M. E. Proctor, M. S. Gradus, K. A. Blair, D. E. Peterson, J. J. Kazmierczak, D. G. Addiss, K. R. Fox, J. B. Rose, and J. P. Davis. 1994. A massive outbreak in Milwaukee of *Cryptosporidium* infection transmitted through the public water supply. N. Engl. J. Med. 331:161–167.

Manitoba Rural Water Quality (MRWQ). 1999. Manure management [Online]. URL: http://www.cwra.org/branches/arts/manitoba/pub1page1.html (Accessed: April 5, 2002).

Millard, P. S., K. F. Gensheimer, D. G. Addiss, D. M. Sosin, G. A. Beckett, A. Houck-Jankoski, and A. Hudson. 1994. An outbreak of cryptosporidiosis from fresh-pressed apple cider. J. Am. Med. Assoc. 272:1592–1596.

Mitscherlich, E., and E. H. Marth. 1984. Microbial survival in the environment. Springer-Verlag, New York. pp. 80, 346, 393, 606, 695, 737–738.

Montville, T. J., and K. R. Matthews. 2001. Principles which influence microbial growth, survival, and death in foods. In: M. P. Doyle, L. R. Beuchat, and T. J. Montville (eds.). Food microbiology: fundamentals and frontiers. 2nd ed. ASM Press, Washington, D. C.

Morgan, G. M., C. Newman, S. R. Palmer, J. B. Allen, W. Shepherd, A. M. Rampling, R. E. Warren, R. J. Gross, S. M. Scotland, and H. R. Smith. 1988. First recognized community outbreak of haemorrhagic colitis due to verotoxin producing *Escherichia coli* O157:H7 in the UK. Epidemiol. Infect. 101:83–91.

Mubiru, D. N., M. S. Coyne, and J. H. Grove. 2000. Mortality of *Escherichia coli* O157:H7 in two soils with different physical and chemical properties. J. Environ. Qual. 29:1821–1825.

Nakazawa, M., M. Akiba, and T. Sameshima. 1999. Swine as a potential reservoir of shiga toxin-producing *Escherichia coli* O157:H7 in Japan. Emerg. Infect. Dis. 5:833–834.

Nelson, H. 1997. The contamination of organic produce by human pathogens in animal manures [Online]. URL: http://eap.mcgill.ca/SFMC_1.htm (Accessed: April 1, 2002).

Olson, M. E., J. Goh, M. Philips, N. Guselle, and T. A. McAllister. 1999. *Giardia* cyst and *Cryptosporidium* oocyst survival in water, soil, and cattle feces. J. Environ. Qual. 28:1991–1996.

Patriquin, D. G. 2000. Reducing risks from *E. coli* O157:H7 on the organic farm [Online]. Summer 2000 issue of Eco-Farm & Garden, magazine of Canadian Organic Growers. URL: http://www.cog.ca/efgsummer2000.htm#ecoli (Accessed: April 3, 2002).

Read, S. C., C. L. Gyles, R. C. Clarke, H. Lior, and S. McEwen. 1990. Prevalence of verocytotoxigenic *Escherichia coli* in ground beef, pork, and chicken in southwestern Ontario. Epidemiol. Infect. 105:11–20.

Riddle, J. A., E. B. Rosen, and L. S. Coody. 1999. American organic standards: guidelines for the organic industry. Organic Trade Association. pp. 31–32.

Rollins, D. M., and R. R. Colwell. 1986. Viable but nonculturable stage of *Campylobacter jejuni* and its role in survival in the natural aquatic environment. Appl. Environ. Microbiol. 52:531–538.

Santo Domingo, J. W., S. Harmon, and J. Bennett. 2000. Survival of *Salmonella* species in river water. Curr. Microbiol. 40:409–417.

Schlech, W. F., P. M. Lavigne, R. A. Bortolussi, A. C. Allen, E. V. Haldene, A. J. Wort, A. W. Hightower, S. E. Johnston, S. H. King, E. S. Nicholls, and C. V. Broome. 1983. Epidemic listeriosis-evidence for transmission by food. N. Engl. J. Med. 308:203–206.

Statistics Canada. 2000. Food consumption in Canada, Part 1 [Online]. URL: http://www.statcan.ca/english/IPS/Data/32-229-XIB.htm (Accessed: October 28, 2002).

Statistics Canada. 2001. Food consumption in Canada, Part II [Online]. URL: http://www.statcan.ca/english/IPS/Data/32-230-XIB.htm (Accessed: October 28, 2002).

Taormina, P. J., L. R. Beuchat, and L. Slutsker. 1999. Infections associated with eating seed sprouts: an international concern. Emerg. Infect. Dis. 5: 626–634.

Terzieva, S. I., and G. A. McFeters. 1991. Survival and injury of *Escherichia coli*, *Campylobacter jejuni*, and *Yersinia enterocolitica* in stream water. Can. J. Microbiol. 37:785–790.

Todd, E. C. D. 1990. Foodborne disease in Canada, 10-year summary 1975–1984. Health Protection Branch, Health and Welfare Canada. Polyscience Publications Inc., Morin Heights, Quebec.

Todd, E. C. D. 1991. Foodborne and waterborne disease in Canada, annual summaries 1985 and 1986. Health Protection Branch, Health and Welfare Canada. Polyscience Publications Inc., Morin Heights, Quebec.

Todd, E. C. D., and C. Harboway. 1994. Foodborne and waterborne disease in Canada, annual summary 1987. Health Protection Branch, Health Canada. Polyscience Publications Inc., Morin Heights, Quebec.

Todd, E. C. D., and P. Chatman. 1996. Foodborne and waterborne disease in Canada, annual summaries 1988 and 1989. Health Protection Branch, Health Canada. Polyscience Publications Inc., Morin Heights, Quebec.

Todd, E. C. D., and P. Chatman. 1997. Foodborne and waterborne disease in Canada, annual summaries 1990 and 1991. Health Protection Branch, Health Canada. Polyscience Publications Inc., Morin Heights, Quebec.

Todd, E. C. D., and P. Chatman. 1998. Foodborne and waterborne disease in Canada, annual summaries 1992 and 1993. Health Protection Branch, Health Canada. Polyscience Publications Inc., Laval, Quebec.

Todd, E. C. D., P. Chatman, and V. Rodrigues. 2000. Annual summaries of foodborne and waterborne disease in Canada, 1994 and 1995, Health Products and Food Branch, Health Canada. Polyscience Publications Inc., Laval, Quebec.

Tschäpe, H., R. Prager, W. Streckel, A. Fruth., E. Tietze, and G. Böhme. 1995. Verotoxinogenic *Citrobacter freundii* associated with severe gastroenteritis and cases of haemolytic uraemic syndrome in a nursery school: green butter as the infection source. Epidemiol. Infect. 114:4410–450.

United States Department of Agriculture (USDA). 1997a. Changes in U.S. swine management practices, 1990–1995 [Online]. National Animal Health Monitoring System. URL: http://www.aphis.usda.gov/vs/ceah/cahm/Swine/sw95chng.pdf (Accessed: April 3, 2002).

USDA. 1997b. Per capita pork consumption—selected countries [Online]. Foreign Agricultural Service. URL: http://www.fas.usda.gov/dlp2/circular/1997/97-03/porkpcap.htm (Accessed: April 3, 2002).

USDA. 1997c. Pork summary-selected countries [Online]. Foreign Agricultural Service. URL: http://www.fas.usda.gov/dlp2/circular/1997/97-03/porksumm.htm (Accessed: April 3, 2002).

USDA. 1999. Per capita pork consumption—selected countries [Online]. Foreign Agricultural Service. URL: http://www.fas.usda.gov/dlp2/circular/1999/99-10lp/porkpcc.pdf (Accessed: April 3, 2002).

USDA. 2000. Pork summary-selected countries [Online]. Foreign Agricultural Service. URL: http://www.fas.usda.gov/dlp/circular/2000/00-03lp/porkpr.pdf (Accessed: April 3, 2002).

USDA. 2001a. Pork summary—selected countries [Online]. Foreign Agricultural Service. URL: http://www.fas.usda.gov/circular/2001/01–03lp/livestock.html (Accessed: April 3, 2002).

USDA. 2001b. Ecology and epidemiology of *Salmonella* and other foodborne pathogens in livestock. National Animal Disease Center. Pre-Harvest Food Safety & Enteric Disease Research Unit [Online]. URL: http://www.nadc.ars.usda.gov/research/fs/epidemiology/ (Accessed: April 1, 2002).

USDA. 2002. Pork summary-selected countries [Online]. Foreign Agricultural Service. URL: http://www.fas.usda.gov/dlp/circular/2002/02-03LP/pk_sum.pdf (Accessed: April 3, 2002).

Valdes-Dapena Vivanco, M. M., and M. M. Adam. 1983. Survival of *Campylobacter jejuni* in different media and faeces at different temperatures and times of preservation. Acta Microbiol. Hung. 30: 69–74.

Wang, G., and M. P. Doyle. 1998. Survival of enterohemorrhagic *Escherichia coli* O157:H7 in water. J. Food Protect. 61:662–667.

Wang, G., T. Zhao, and M. P. Doyle. 1996. Fate of Enterohemorrhagic *Escherichia coli* O157:H7 in bovine feces. Appl. Environ. Microbiol. 62:2567–2570.

Weichart, D. H. 1999. Stability and survival of VBNC cells-conceptual and practical implications. In: C. R. Bell, M. Brylinsky, and P. Johnson-Green (eds.). Microbial Biosystems: New frontiers. Proceedings, 8th International Symposium on Microbial Ecology, Atlantic Canada Society for Microbial Ecology, Halifax.

Zibilske, L. M., and R. W. Weaver. 1978. Effect of environmental factors on survival of *Salmonella typhimurium* in soil. J. Environ. Qual. 7:593–597.

CHAPTER 3

Pork Production and Human Health in the Major Pork Producing Countries

SUMMARY

The present situation regarding pork production and its potential effect on human enteric illness in Denmark, the Netherlands, Taiwan, the United States, and Canada are discussed. Among these countries, the Netherlands has the highest hog density, followed by Taiwan, Denmark, North Carolina (U.S.), and Quebec (Canada). Denmark has the highest per capita pork consumption, followed by the Netherlands, Taiwan, Canada, and the U.S. In terms of human enteric illness, Denmark has the highest *Salmonella*, *Campylobacter*, and *Yersinia* cases per 100,000 population, while Canada has the highest *Escherichia coli* O157:H7 cases. Foodborne illness data for Taiwan is not available for comparison. The prevalence of *Salmonella* and *Campylobacter* in pigs is highest in the U.S., while the prevalence of *Yersinia* is highest in Canadian swine. Pork-associated human illness in Denmark and the Netherlands seems to be affected by pork consumption level, pathogen prevalence in pigs, and hygiene at slaughter. Although pathogens are highly prevalent in Canadian and American swine, the cases of human enteric illness due to pork seem unrelated. It is believed that lower pork consumption and the tendency for thorough cooking of pork in these two countries may have influenced the low rate of pork-associated human illness. Despite an increase in pork production in the last decade, pork is not a major source of human enteric illness in Canada.

DENMARK

Denmark has a population of 5.35 million and an area of 44,000 km^2 (Anon 2002), resulting in a population density of 121 people per km^2. It had the third

Table 3.1. Hog densities in selected countries or regions in 2000.[a]

Country or region	Hog density (animals /km² or mile²)	
	Kilometre²	Mile²
Netherlands	1,484.6	3,845.3
Taiwan	812.9	2,105.5
Denmark	503.8	1,304.8
South Korea	378.8	981.1
Japan	245.2	635.1
Germany	220.3	570.5
Spain	144.8	375.0
France	87.7	227.0
United States—North Carolina	465.1	1,204.7
United States—Iowa	145.0	375.7
United States—Minnesota	62.7	162.3
Canada—Quebec	213.9	554.3
Canada—Ontario	90.6	234.6
Canada—Manitoba	**38.5**	**99.8**
Canada—Alberta	16.2	41.9

[a]Modified from SAF (2001).

highest hog density in the world in 2000 (504 animals per km²) which results in four hogs for each person per unit of land area (Table 3.1). Hog production in the country has gradually increased in the past decade, with production at roughly 22 million finisher pigs in 2000 (Table 3.2). In fact, Denmark has always been one of the world's leading pork exporters (Table 3.3), exporting about 80% of its production (Anon 1997). Denmark also has the highest per capita pork consumption in the world, with consumption at least three times higher than that of beef (Table 3.4).

The above observations may explain why pork is considered a major source of human enteric illness in Denmark. Pork is believed to be responsible for most of the human infections caused by *Salmonella enterica* serovar Typhimurium and *Yersinia enterocolitica*. In fact, the majority (about 70%) of the

Table 3.2. Hog production in Denmark, 1990–2000.[a]

Year	×1000 head
1990	16,427
1992	18,559
1994	20,760
1996	20,530
1998	22,873
1999	22,534
2000	22,411

[a]Modified from DANMAP (2000).

Table 3.3. World's top 10 pork
exporters in 2000.[a]

Country	% of world exports
Canada	**23.0**
United States	18.1
Denmark	17.2
France	6.4
Poland	5.0
Germany	4.6
Netherlands	4.0
China	3.4
Austria	1.8
Hong Kong	1.7

[a]Modified from USDA (2001a).

S. Typhimurium infections acquired by humans domestically was attributed to pork (Hald & Wegener 1999). *S.* Typhimurium definitive type (DT) 12, which predominates in pig herds in Denmark, also occurred most frequently in humans. In 1996, thirty different serotypes of *Salmonella enterica* were isolated from 6.2% of Danish finisher pigs (n = 13,468), with *S.* Typhimurium as the predominant serotype (61%). Salmonellosis is one of the major zoonotic diseases in Denmark (Table 3.5). A mandatory surveillance program for monitoring *Salmonella* in pork at the slaughterhouse level and serological monitoring of slaughter pig herds were initiated in 1993 and 1995, respectively (Hald & Andersen 2001). Before the control program was initiated, the level of *Salmonella* in fresh pork produced in Denmark was approximately 3% (Nielsen & Wegener 1997). In 2000, the *Salmonella* level was reduced to 0.8% of pork cuts at the slaughterhouse and 1.1% of retail pork samples (Anon 2001). Concomitantly, the number of human *S.* Typhimurium infections has also gradually decreased (Table 3.5).

Table 3.4. Per capita pork, beef and veal consumption for selected
countries in 2000.[a]

Country	Pork consumption[b]	Beef & veal consumption[b]
Denmark	63.1	22.7
Germany	56.8	14.3
Netherlands	44.0	19.0
Taiwan	43.0	3.9
France	37.6	25.6
Canada	**31.1**	**32.1**
United States	31.0	45.4
Korea	22.5	11.5
Japan	17.2	12.0

[a]Modified from USDA (2001b, c).
[b]Per capita based on carcass weight equivalent (kg).

Table 3.5. Hog production, per capita pork consumption, human illness, and foodborne disease in Denmark.[a]

Year	Production (tonnes)	Per capita consumption (kg)	Laboratory notified cases of diseases[b]						Foodborne disease[d]	
			Salmonella	S. Typhimurium	Campylobacter	Yersinia	VTEC	O157[c]	Outbreaks	Cases
1985						1,512 (29.0)				
1987			2,619	1,142	1,518	1,143	7	2		
1988			3,495	1,826	1,445	1,015	3	1		
1989			2,601	1,044	1,432	879	8	1		
1990		64.2	2,112	728	1,367	967	32	2		
1991			2,238	700	1,261	929	49	6		
1992			3,379	1,289	1,129	909	10	6		
1993			3,802	1,193	1,776	710	4	1		
1994			4,276	1,363	2,196	643	10	3		
1995	1,494,000	64.1	3,654 (70.2)[e]	848	2,601	779 (15.0)	2	2	31	186
1996	1,493,700	64.6	3,259	907	2,973	532 (9.0)	5	3	40	542
1997	1,520,600	57.1	5,015 (95.0)	841 (15.9)	2,666 (50.5)	430 (8.1)	33 (0.6)	12 (0.2)	44	770
1998	1,629,300	63.1	3,880 (73.3)	678 (12.8)	3,372 (63.7)	464 (8.8)	34 (0.6)	6 (0.1)	46	610
1999	1,641,800	65.8	3,268 (61.5)	584 (11.0)	4,164 (78.4)	339 (6.4)	51 (1.0)	18 (0.2)	86	1785
2000	1,624,500	63.1	2,308 (43.3)	436 (8.1)	4,386 (82.3)	265 (5.0)	60 (1.1)	18 (0.3)	77	979
2001	1,705,000		2,918 (54.5)	589 (11.0)	4,620 (86.4)	286 (5.3)	92 (1.7)	24 (0.4)		

[a]Modified from WHO (1999), Anon (1998, 1999, 2000, 2001, 2002).
[b]Laboratory notification system, where the diagnostic laboratories report on all patients with a culture positive for Salmonella, Campylobacter, Yersinia, and VTEC, all culture confirmed cases of infection caused by these agents must be reported.
[c]Includes all O157 serotypes with or without H7 flagellar antigen.
[d]Food control system, where individuals who experienced food poisoning reported the incidents, these reports as well as the result of the outbreak investigations are collated. Outbreaks of foodborne diseases must be reported if there is well-founded evidence that a specific foodstuff or a specific meal caused the illness. Outbreaks are mentioned if the causative agent has been found in either patients or food. The numbers of cases refer to the total number of patients who have had symptoms that could have been caused by incriminated food and are not based on laboratory confirmed results.
[e]Numbers in parenthesis are cases per 100,000 inhabitants.

In fact, seasonal variation of *Salmonella* prevalence in pork and the human incidence were found to follow a very similar course (Hald & Andersen 2001). The declining trend in *Salmonella* prevalence in slaughter pigs and pork are primarily believed to be a result of the integrated control efforts implemented at the herd and slaughterhouse level. The control program is also presumed to be responsible for at least a part of the reduction observed in human *S.* Typhimurium cases (Table 3.5). This presumption is supported by the results of Hald & Wegener (1999) which indicated that the number of human cases caused by pork-associated *Salmonella* types has been gradually reduced since 1996. Among other reasons, it may partially be an effect of a greater awareness of the risk among the public.

The majority of Danish slaughter pig herds is assumed to harbour *Y. enterocolitica* serotype O:3. A serological survey in 1993 showed that 90% of herds and 75% of slaughter pigs were contaminated (Anon 2001). A within-herd prevalence of 80% was also reported. As a result, pork is believed to be the primary source of human yersiniosis in the country. The number of human infections with *Y. enterocolitica* peaked in 1985 when 1,512 cases were identified (29 cases per 100,000) but this gradually declined to 265 cases in 2000 (5 cases per 100,000), with the majority (97.3%) of cases caused by serotype O:3 (Table 3.5) (Nielsen & Wegener 1997; Anon 2001). The reduction in human yersiniosis is believed to be a result of improved hygiene at slaughter, that is, reduction of fecal contamination of the carcasses (Nielsen & Wegener 1997). Nonetheless, the decline of human yersiniosis cases coincides with the finding that only 13% of 316 pigs examined in 2000 were positive for *Y. enterocolitica* O:3 (Anon 2001). However, an investigation of raw retail pork in 2000 still revealed that 6.3% of samples contained the pathogen.

Although *Listeria monocytogenes* can be detected at relatively high rates in pigs and pork in Denmark (Nielsen & Wegener 1997), only sporadic cases of human listeriosis have been observed. These occurred at a relatively low rate of 0.5 to 0.8 cases per 100,000 inhabitants between 1986 and 2000 (Anon 2001). This is consistent with a U.S. rate of one case per 100,000 persons (Potter et al. 2002). A low level of strain virulence for humans is apparent. Swine are also not considered to be a source for *Campylobacter jejuni* or VTEC in this country, although human infections due to these pathogens are on the rise in recent years (Table 3.5). Despite a gradual increase in hog production, domestic pork consumption did not vary much between 1995 and 2000 (Table 3.5) which is understandable as most of the additional production was directed for export.

THE NETHERLANDS

The Netherlands has a population of 15.8 million persons and an area of 41,532 km², resulting in a population density of 380 people per km² (WHO 1999). It has the highest ratio of hogs to land area in the world, with a pig density of 1,485 animals per km² (Table 3.1). Just as in Denmark, this yields about four pigs for each person per unit of land. The Netherlands is one of the world's top ten pork exporters (Table 3.3), with exports representing about 65% and 75% of its pork and pig production, respectively (Anon 1997). Per capita pork consumption, which is more than twice that of beef, is among the world's highest (Table 3.4).

Meat and meat products are responsible for about 10–15% of foodborne illness in the Netherlands. Occurrence of outbreaks due to these products has fluctuated in the past decade (Table 3.6). Pork is the most frequently consumed type of meat in this country, followed by poultry (Heuvelink et al. 1999a). Consumption of pork has been associated with salmonellosis. In 1997, 43% of *Salmonella enterica* serovar Typhimurium infections in the Netherlands originated from the consumption of pork (Hald & Wegener 1999). Other dominating serotypes among the pork-related cases include *S.* Infantis, *S.* Panama, *S.* Derby, and *S.* Livingstone. Pork was assumed to be responsible for 14–19% (2.4–3.2 cases per 100,000) of all *Salmonella* cases in 1997. In fact, the *Salmonella* serotypes associated with pigs and pork in the country over the past decade have accounted for an average of 15% of identified cases of human salmonellosis. It is also estimated that the number of cases of salmonellosis associated with the production and consumption of pork to be about 10,000 per year (Berends et al. 1998), although the notified cases of human salmonellosis are only about one fifth of the number (Table 3.6). *Salmonella* contamination appears to be common in pigs in the Netherlands. The prevalence of *Salmonella* in finisher pigs can be as high as 44% (van der Wolf et al. 2001) while a herd prevalence of 23% has been reported (van der Wolf et al. 1999). It is believed that about two thirds of all Dutch pig farms are more or less permanently infected (Berends et al. 1996). At infected farms, the probability that *Salmonella*-free pigs will become infected is about 85–90%. Further study revealed that pigs which carry *Salmonella* are three to four times more likely to end up as a positive carcass (Berends et al. 1997). It was found that 70% of all carcass contamination results from the animals themselves being carriers, while 30% is due to cross contamination from other carrier animals. It is also estimated that between 5–30% of the hog carcasses produced in the Netherlands contain *Salmonella* (Berends et al. 1997).

In the Netherlands, pigs are considered to be the only known source for *Yersinia enterocolitica* O:3 and O:9, which are also the main serotypes isolated from patients with yersiniosis in the country (de Boer & Nouws 1991; de Boer et al. 1986). In the earlier study, de Boer et al. (1986) isolated non-pathogenic *Yersinia* strains from a great variety of foods while only pathogenic serotypes were isolated from porcine tonsils. In the later study (de Boer & Nouws 1991), pathogenic serotypes O:3, O:9, and O:5,27 were isolated from 42% of porcine tonsils, 20% of porcine tongues, 17% of rectal swabs, and 1% of pork samples. However, none was isolated from pig carcasses. Thus, it was thought that contamination of carcasses during the slaughtering process with *Yersinia* from fecal material or from the tonsillary region does not frequently occur. This may explain the low contamination rate in pork. Human yersiniosis in the Netherlands was under surveillance only until 1996, with incidence rates between 0.6 to 0.9 cases per 100,000 population between 1993 and 1996 (Table 3.6). It is

Table 3.6. Hog production, per capita consumption, and foodborne diseases in the Netherlands.

Year	Production (×1000 head)[a]	Per capita consumption (kg)[b]	Outbreaks from meat and meat products[c]	Notified cases of foodborne diseases[c,d]			
				Salmonella	Campylobacter	VTEC O157[e]	Yersinia
1993		53.8	31	2804 (17.7)			111 (0.7)
1994		43.7	47	2885 (18.3)			136 (0.9)
1995		44.2	45	2826 (17.9)	2871 (18.2)[g]		111 (0.7)
1996	14,419	44.3	64	2889 (18.3)	3741 (23.7)	10 (0.1)	89 (0.6)
1997	15,189	43.1	98	2556 (16.2)	3646 (23.1)	29 (0.2)	
1998	13,446	44.2	38	2266 (14.3)	3398 (21.5)	31 (0.2)	
1999	13,567	44.2	94 (479)[f]	2128	3160	36	
2000	13,118	44.0	87 (439)	2059	3362	43	

[a]Statistics Netherlands (2001).
[b]USDA (1998, 2001b).
[c]WHO (1999, 2002).
[d]Laboratory-based surveillance for infectious diseases, in which 15 regional public health laboratories (16 for *Salmonella*) weekly forward the number of positive test results and the number of fecal samples tested. Numbers in parenthesis are incidence rates per 100,000 population.
[e]Includes all serotypes of O157 with or without H7 flagellar.
[f]Numbers in parenthesis in this column are numbers of cases in outbreaks.
[g]Surveillance started in April, 1995.

unclear why *Yersinia* cases are no longer included in the statutory notification system. Perhaps it is related to the low incidence rate of yersiniosis in the country.

The role that pigs play in the epidemiology of human *E. coli* O157 (with or without H7 flagellar antigen) infection in the Netherlands is not well understood. Ever since VTEC O157 was included in the statutory notification system in 1996, the number of notified cases of human O157 infection has been steadily rising (Table 3.6). Potentially pathogenic *E. coli* O157 strains were isolated from 1.4% of Dutch slaughter pigs in 1997 and 1998 (Heuvelink et al. 1999a). An examination of retail meats in the Netherlands revealed *E. coli* O157 contamination in 1.3% of minced pork, 0.5% of minced mixed pork and beef products, 0.3% of ready-to-eat dry fermented pork sausage, and 0.3% of other raw pork products (Heuvelink et al. 1999b).

Although *Campylobacter* is a leading cause of foodborne illness in the Netherlands (Table 3.6), pork has not been implicated as the vehicle of transmission. While per capita pork consumption has been stable for the past decade, hog production and cases of human salmonellosis have declined a little in this country (Table 3.6).

TAIWAN

Prior to the foot and mouth disease outbreak in March 1997, Taiwan was among the top 15 producers of pork and pork products worldwide (USDA 2001a). It was second only to the United States in pork exports in 1996 (contributing 15.2% of world exports). Nearly 40% of the hogs raised in the country were used for the export market (Wilson & Tuszynski 1997). Ever since the outbreak hit the industry pork export has been banned. Nevertheless, pig density in this country is still among the highest, and was second only to the Netherlands in 2000 (Table 3.1). The country is comprised of two thirds mountains, while the other third has the highest density of pig farms in the world (Wayt 2001). Per capita pork consumption in this country is also among the world's highest, with a consumption level ten times more than that of beef (Table 3.4). Pork is the preferred meat in Taiwan and it accounts for 60% of total meat consumption (Wayt 2001).

Early statistics in Taiwan showed that red meat and poultry together were responsible for only about 5% of all foodborne illnesses, of which more than 50% were of unknown etiologies (Chiou et al. 1991). Unlike North America and Europe, the leading cause of foodborne illness outbreaks in Taiwan is *Vibrio parahaemolyticus* which is transmitted mainly by consumption of seafood (about 17% of all foodborne illness). Between 1986 and 1995, *V. parahaemolyticus* was responsible for 35% and 38% of foodborne bacterial outbreaks and cases, respectively. This was followed by *Staphylococcus aureus* (30% and 28%), *Bacillus cereus* (18% and 20%), pathogenic *E. coli* (6.5% and 6.0%), *Salmonella* (5.6% and 4.5%), *Clostridium botulinum* (1.8% and <0.1%), and others (1.4% and 1.5%) (Pan et al. 1997). More recent data (between 1996 and 1999) showed a dramatic increase of foodborne disease outbreaks caused by

Table 3.7. Hog production, per capita pork consumption, and
foodborne diseases in Taiwan.

Year	Production 1000 metric tons[a]	Per capita pork consumption (kg)[b]	*Salmonella* outbreaks[c]
1992	1,113		
1993	1,135	40.8	
1994	1,204	41.4	
1995	1,233	40.3	8
1996	1,269	42.1	10
1997	1,030	40.0	7
1998	892	44.7	8
1999	822	43.3	11
2000	921	43.0	
2001	910	43.2	

[a]USDA (1997, 2000, 2001a, 2002a).
[b]USDA (1998, 2001b).
[c]Chiou et al. (2000).

V. parahaemolyticus (66% of all foodborne outbreaks). This was followed by *Salmonella* (5.2%), *S. aureus* (1.8%), *B. cereus* (0.7%), and others (chemical and unknown etiologies, 26%) (Chiou et al. 2000). No cases of foodborne *Campylobacter*, *Yersinia*, and VTEC O157 have been reported. This is most probably because the testing for these organisms is not included in the surveillance program in this country. It must be kept in mind that about 26–50% of foodborne diseases involved unidentified vehicles.

Although *Salmonella* is not a major cause of foodborne illness in Taiwan, recent research has linked the occurrence of antibiotic resistant *Salmonella* strains in humans to pigs. A study by Chiu et al. (2002) showed a rapid increase in fluoroquinolone resistance in *Salmonella enterica* serotype Choleraesuis in humans over the previous two years and indicated that the resistant strain was spread from pigs. Unlike multidrug-resistant *S. enterica* Typhimurium, Hadar, and Enteritidis, which usually cause gastroenteritis, *S. enterica* Choleraesuis causes systemic infections in humans. In swine, the organism causes septicemia. Although *Campylobacter* has not been associated with foodborne disease outbreaks in Taiwan, the organism has been isolated from patients with gastroenteritis. It was isolated from 5.6% of patients in the Taipei metropolitan area in 1981 and 2.5% in central Taiwan between 1994 and 1996 (Tang et al. 1981 & Lin et al. 1998 cited by Lu et al. 2000).

One reason that probably contributes to the low rate of meat-associated foodborne illness in Taiwan is that meat, in particular pork, tends to be thoroughly cooked in Chinese cuisine. Per capita pork consumption seems fairly stable over the past 10 years with a slight increase after 1997 (Table 3.7). It is of interest that this appears to be in a reverse relationship with hog production which has steadily increased until it peaked in 1996, just before the foot and mouth disease outbreak. Due to insufficient data from foodborne illness outbreaks in Taiwan, an accurate evaluation of the relationship between pork production and human health cannot be made.

Table 3.8. Foodborne illness outbreaks from
USDA-regulated pork products, 1990–1998.[a]

Year	Outbreaks	Cases
1990	6	246
1991	4	231
1992	3	120
1993	5	241
1994	8	163
1995	5	331
1996	5	187
1997	4	835
1998	4	326
Total	44	2,680

[a]Modified from CSPI (2001).

THE UNITED STATES

In the United States, consumption of pork is lower than that of beef and
veal (Table 3.4). From 1992 to 2000, per capita pork consumption in the U.S. remained fairly stable despite a gradual increase in pork production (Table 2.4). As
with some other pork producing countries, this increase in pork production was
used to supply export markets. The U.S. was the second largest pork exporter
in 2000 (Table 3.3). It is estimated that 14.6% of all known causes of foodborne
illness outbreaks in the U.S. are attributed to pork consumption (USDA 2001d).
The U.S. statistics from 1973 to 1992 showed that there has been a drop in
the total number of foodborne disease outbreaks attributed to pork (Table 2.3).
However the total number of cases in these outbreaks was not available. The
numbers of outbreaks and cases of USDA-regulated pork products (including
ham) from 1990 to 1998 (Table 3.8) fluctuated and thus no trend can be observed.
During this period pork was responsible for about 15% of all meat and poultry
outbreaks, with the highest average of cases per outbreak in any one year being,
61 for pork, 45 for beef, and 54 for poultry. The U.S. hog carcass baseline data for
Salmonella (8.7%), *Campylobacter* (31.5%), and *Listeria monocytogenes* (7.4%)
suggested the widespread distribution of human pathogens in U.S. pork (FSIS
1996). *Salmonella* levels were selected as the performance standard for the national HACCP program for slaughter plants and establishments that produce raw
ground products. In 2001, the *Salmonella* prevalence in swine was reduced to
1.6% in large plants (those with 500 or more employees), 3.5% in small plants
(those with 10 or more but fewer than 500 employees), and 4.4% in very small
plants (those with fewer than 10 employees) (FSIS 2002). Several studies have
determined the prevalence of foodborne pathogens in retail pork. Duffy et al.
(2001) reported incidences of *Salmonella* (9.6%), *L. monocytogenes* (19.8%),
Y. enterocolitica (15.6%), and *Campylobacter* (1.3%) on retail pork products
from six cities. The unusual high occurrence rates of *L. monocytogenes* and
Y. enterocolitica in these retail products deserve attention. Another study by

Table 3.9. Hog slaughter by state in
the U.S. in 2001.[a]

State	1,000 head
Iowa	27,371.5
North Carolina	9,888.2
Illinois	9,490.8
Minnesota	8,465.1
Nebraska	6,681.4
Indiana	6,558.7
Oklahoma	4,500.5
South Dakota	4,082.8

[a]Modified from USDA (2002b).

Zhao et al. (2001) showed a prevalence of *Salmonella* (3.3%), *Campylobacter* (1.7%), and *E. coli* (16.3%) in retail pork from Washington, D.C. The low rate of *Salmonella* contamination of pork products reported in the Washington study agrees with the recently released USDA performance standard surveillance data (see above).

The major hog producing states in the U.S. are shown in Table 3.9. Iowa, North Carolina, and Minnesota, which produce the highest number of hogs in the nation, are also the states with the highest hog densities (Table 3.1). Iowa alone had 5 hogs for every person in 2001 which is even higher than Denmark and the Netherlands (IDALS 2001). Reportable diseases caused by *Salmonella, Campylobacter*, and *E. coli* O157:H7 in Iowa are presented in Table 3.10. Numbers of infections caused by all three pathogens have been increasing since they were first reported. By decade, salmonellosis cases in 1991–2000 were at least

Table 3.10. Reportable diseases (number of cases) caused by
Salmonella, Campylobacter, and *E. coli* O157:H7 in Iowa.[a]

Year	Salmonella	Campylobacter	E. coli O157:H7
1951–1960	288		
1961–1970	1,096		
1971–1980	2,069		
1981–1990	2,754	3,566	
1991–2000	3,361	3,624	804
1991	304	333	15
1992	339	260	20
1993	242	292	27
1994	404	280	54
1995	433	274	64
1996	335	339	123
1997	296	425	114
1998	375	455	93
1999	260	467	114
2000	373	499	180

[a]Modified from IDPH (2001).

Table 3.11. Top *Salmonella* serovars in the U.S. and Iowa—1995.[a]

U.S. top ten *Salmonella* serovars (human)	Source of *Salmonella* serovars isolated in Iowa			
	Human	Swine	Cattle	Chicken
Enteritidis	Typhimurium	Cholerasuis	Typhimurium	Enteritidis
Typhimurium	Enteritidis	Typhimurium	Dublin	Heidelberg
Newport	Agona	Anatum	Anatum	Kentucky
Heidelberg	Newport	Agona	Kentucky	Hadar
Hadar	Infantis	Heidelberg	Montevideo	Typhimurium
Javiana	Heidelberg	Brandenburg	Muenster	Agona
Muenchen	Hadar	Infantis	Cerro	Montevideo
Montevideo	Braenderup	Enteritidis	Enteritidis	Senftenberg
Agona	Montevideo	Bredeney	Mbandaka	Infantis
Thompson	Muenchen		Infantis	Schwartzongrund
			Agona	Mbandaka
			Senftenberg	Cerro
			Hadar	Anatum
			Heidelberg	Bredeney
			Bredeney	

[a]Modified from Moyer (1997a,b).

10 times more than in 1951–1960, and *Campylobacter* cases were also higher in 1991–2000 than the decade before. Infections caused by *E. coli* O157:H7 have also been steadily rising since its first reportable year in 1991. This phenomenon is most probably a result of increased diagnostic awareness and an enhanced surveillance program in Iowa. Nonetheless, the number of cases due to these pathogens showed no trend toward decreasing during the last decade and deserves some attention from public health officials. The most frequently isolated *Salmonella* serovars from humans and livestock in Iowa are presented in Table 3.11. Five serovars, namely Typhimurium, Enteritidis, Agona, Newport, and Infantis, accounted for 71% of reported human cases in 1995 (Moyer 1997b). In the list of top ten Iowa *Salmonella* serovars from humans, all serovars except Newport, Braenderup, and Muenchen were also frequently isolated from livestock (Table 3.11). Comparison of the swine and human serovars indicate that serovars Typhimurium, Agona, Heidelberg, Infantis, and Enteritidis appear on both lists. This is probably a result of high agricultural activities in this state. Ninety-one percent of the land area in Iowa is used for agriculture, which is the highest percentage in the U.S. (IDALS 2001). Besides ranking as the number one state in pork production, Iowa was second in egg production, and third in total red meat production in the U.S. in 2001.

It has been said that there is no reliable information regarding the prevalence of *Salmonella* in swine (Fedorka-Cray et al. 1997) and there is no best protocol to assess *Salmonella* prevalence in swine (Hurd et al. 2001). This is probably because *Salmonella* contamination in hogs varies greatly among different intestinal sites. While *Salmonella* was negative in fecal samples, positive

Table 3.12. Incidence per 100,000 population of diagnosed infections for four pathogens by year and organism—Foodborne Diseases Active Surveillance Network, United States.[a]

Pathogen	Original 5 sites					8 sites	9 sites[b]	National health objective for 2010
	1996	1997	1998	1999	2000	2000	2001	
Campylobacter	23.5	25.2	21.4	17.5	20.1	15.7	13.8	12.3
Salmonella	14.5	13.6	12.3	13.6	12.0	14.4	15.1	6.8
E. coli O157:H7	2.7	2.3	2.8	2.1	2.9	2.1	1.6	1.0
Y. enterocolitica	1.0	0.9	1.0	0.8	0.5	0.4	0.4	NA[c]

[a]Modified from CDC (2001, 2002).
[b]13% of the US population.
[c]Not applicable.

results (17–49%) were obtained from ileocecal lymph nodes, cecal contents or colons of Iowa pigs (McKean et al. 2000). Therefore it is imperative to prevent *Salmonella* contamination of hog carcasses from all these sources at the slaughterhouse level. Perhaps another reason for ambiguity is the difference in methodology used. One study on Iowa swine farms found that at slaughter, 52% of swine were serologically positive for *Salmonella* antibodies, while only 9% were positive by culture (Fedorka-Cray et al. 1997). Prevalence of the organism in pigs appears to fluctuate over time on the farm. It was revealed that about 90% of pigs tested were positive by serology at one week of age. This decreased to 15% by nine weeks, and rose to 52% at slaughter (Fedorka-Cray et al. 1997). All these factors have made an accurate determination of *Salmonella* prevalence in swine difficult. Among *Salmonella* isolates from Iowa swine, serovars Choleraesuis and Typhimurium are most frequently encountered (Table 3.11). In swine, *S.* Choleraesuis is associated with septicemia while *S.* Typhimurium is associated with enterocolitis; both are known to infect humans (Fedorka-Cray et al. 1996). Serotyping for *Salmonella* isolates from Iowa swine in 2000–2001 identified *S.* Typhimurium in 32% of all swine isolates, and 80% of *S.* Typhimurium isolates were *S.* Typhimurium var. Copenhagen (USAHA 2001). *S.* Typhimurium and *S.* Typhimurium var. Copenhagen were also the most common serotypes from North Carolina swine herds (McKean et al. 2000). Although pork is a major source of foodborne salmonellosis throughout the world (see Denmark and Netherlands), more cases of salmonellosis in the U.S. are linked to beef and poultry than pork, partially because the fear of *Trichinella* has encouraged people to cook pork more thoroughly.

FoodNet surveillance data, which covered 13% of the U.S. population in 2001, yielded incidence rates per 100,000 population for a number of human pathogens from 1996 to 2001 (Table 3.12). *Salmonella* incidence rates initially seemed to decrease in the five original sites from 1996 to 2000, but in 2001 which included 9 sites, the rate rose to 15.1 per 100,000 population. The nine site incidence rate is a better estimate of the national data. This rate is more than twice that set for the national health objective for 2010, which appears to be the most challenging objective for all pathogens. Clearly, more effort is needed in order to reach the *Salmonella* objective.

Campylobacter is the leading cause of foodborne infection in the U.S. Most *Campylobacter* infections are caused by *C. jejuni*. *C. coli*, a less severe pathogen than *C. jejuni*, is found more frequently in pigs and pork (Wesley et al. 1998). Researchers in Iowa and North Carolina identified *C. coli* in about 70% of 1,300 market-weight hogs, while less than 1% of these healthy hogs harbored *C. jejuni*. *C. coli* contamination appears to be ubiquitous in pigs in these two states. Research on swine farms found *C. coli* in 90.2 and 89% of nursery pigs, 96 and 89% of grower pigs, 93.8 and 88% of finisher pigs in Iowa and North Carolina, respectively (Wesley et al. 1998). At slaughter the organism was detected on 9 and 25.2% of hog carcasses in Iowa and North Carolina, respectively. For Iowa hogs, 83% of ileocaecal lymph nodes also harbored *Campylobacter coli*. These results were consistent with those from a later study by McKean et al. (2000), who reported fecal *Campylobacter coli* isolation of 80–100% from all stages of production in Iowa and North Carolina swine production units and isolation rates of 35–88% from lymph nodes of Iowa hogs. Given the high prevalence of *Campylobacter coli* in pigs at all production stages, it seems it would be difficult to reduce contamination levels with on-farm interventions under current production practices. FoodNet data from 1996 to 2001 showed a gradual decrease in the incidence rates for *Campylobacter* infection, and with continuing effort it may be possible to reach the *Campylobacter* objective in year 2010 (Table 3.12).

Clinically healthy pigs can harbor *Y. enterocolitica* and are regarded as a significant reservoir for human infection. The pathogen has been isolated from tonsils, tongue, and rectal swabs of healthy pigs (de Boer & Nouws 1991; de Boer et al. 1986; Funk et al. 1998). The first reported outbreak of *Y. enterocolitica* O:3 infections in the U.S. occurred in 1988 in Atlanta, where indirect exposure to contaminated raw pork intestines (household preparation of chitterlings) was identified as the primary mode of transmission (Lee et al. 1991). A study from Iowa State University in 1997, which monitored 100 pigs over 7 months on a large farm, found *Y. enterocolitica* in 7.4% of all fecal samples (ISU 1997). In a later study, *Y. enterocolitica* was detected in 8.8% (5 of 57) of North Carolina hogs but it was not found in rectal, tonsilar or carcass swabs collected from Iowa hogs (Wesley et al. 1998). This may indicate the intermittent carriage of the organism in swine. It may also reflect the problems of reproducibility on repeated sampling which has been documented for *Salmonella* in swine (Isaacson 2000). A large study in Illinois reported 95 of 103 lots of market swine contained at least one pig infected with *Y. enterocolitica*, and 29 lots contained at least one pig carrying pathogenic *Y. enterocolitica* (Funk et al. 1998). In addition, of the 107 pathogenic *Y. enterocolitica* isolates found, 89.7% and 3.7% were pathogenic serotype O:5 and O:3, respectively. The incidence rates for *Y. enterocolitica* according to FoodNet data appear to be slowly decreasing from 1996 to 2001 (Table 3.12). However, FoodNet surveillance sites only include one major hog producing state, i.e. Minnesota. Therefore, it is probable that the *Yersinia* problem in the nation is underestimated in FoodNet data.

CANADA

In Canada, per capita pork consumption is almost the same as that of beef (Table 3.4). Incidents (outbreaks) and cases of pork-associated human enteric illness in Canada, from 1975 to 1995 (the latest statistics available), are presented in Figs 2.1 and 2.2. There is a trend toward a decrease in terms of both total incidents and cases, according to the data. Per capita disappearance of pork in Canada, from 1982 to 2000, did not vary much (Fig. 2.3), despite an apparent increase in national hog production (Fig. 2.4) over the same period of time. There appears to be no relationship among human enteric illness from pork, per capita pork disappearance, and total hog production in Canada. Canada exports the largest amount of pork of all producing countries, contributing 23% of world exports in year 2000 (Table 3.3). The proportion of production exported rose from 21% in 1982 to 40% by the year 2000 (Statistics Canada 2000, 2001). The increase in national hog production has supported increased hog exports while domestic pork consumption has remained stable. While hog production in Manitoba doubled since 1992 (Table 2.2), this change has not affected rates of human illness from pork consumption. Available data indicate that the safety of pork (as a vehicle for foodborne illness) has remained unchanged or improved during this period.

The major hog producing provinces in Canada are Quebec, Ontario, Manitoba, and Alberta (Table 2.2). These provinces also have the highest hog densities in the nation (Table 3.1). Quebec has a human population of 7.41 million and an area of 1,365,128 km^2, resulting in a population density of 5 persons per km^2. It has the highest hog density in Canada, which is 214 animals per km^2, thus resulting in 42 animals for every person in this province. It has the highest ratio of hogs to persons in Canada, followed by Manitoba (19 animals for one person), Ontario (7), and Alberta (4). Manitoba ranks 3rd in pork production and has the second highest ratio of hogs to persons in the nation. Although Canadian provinces have a higher number of hogs per person than Denmark and Holland, both of which have 4 hogs per person, it does not seem to have caused any increase in hog- or pork-associated human illness in Canada. On the other hand, hog densities in Denmark and Holland are 2 to 7 times higher than those of Canadian provinces. As cases of pork-associated human illness are higher in these two countries than in Canada, it appears that hog density has a positive relationship with human illness, but the number of hogs per person does not. Perhaps this should not be surprising given that Canada is, by comparison, sparsely populated by either species.

Regional distribution of foodborne disease outbreaks and cases in Canada during 1980 to 1995 is shown in Table 3.13. During this period, Ontario, Quebec, and British Columbia had the highest number of foodborne illnesses. Foodborne

Table 3.13. Regional distribution of foodborne outbreaks and cases in Canada, 1980–1995.[a]

Year		BC	AB	SK	MB	ON	QC	NB	NS	PEI	NF	NT	YK	CANADA
							Province/Territory							
1980	Outbreaks	152	70	20	24	252	69	6	16	2	6	2	1	621
	Cases[b]	840	827	337	237	2119	1466	638	127	19	398	4	60	7122
	Incidence rate[c]	31.5	38.6	35.1	23.1	24.7	23.0	91.7	8.9	15.5	70.4	15.0	269.1	29.6
1985	Outbreaks	153	44	35	33	365	102	4	17	1	2	5	—	761
	Cases	816	590	150	236	2875	1212	12	452	200	27	39	—	6609
	Incidence rate	28.4	25.1	14.9	22.2	31.9	18.6	1.7	51.9	158.7	4.7	75.1	—	26.3
1990	Outbreaks	144	17	34	8	366	216	11	8	—	2	4	—	810
	Cases	898	104	392	22	2350	1960	176	27	—	5	93	—	6027
	Incidence rate	27	4	39	2	23	28	24	3	—	1	157	—	22
1991	Outbreaks	147	7	29	14	360	228	2	3	—	1	—	—	791
	Cases	881	106	173	68	2718	2390	49	18	—	3	2	—	6408
	Incidence rate	26	4	17	6	26	34	7	2	—	1	3	—	23
1992	Outbreaks	128	21	30	1	371	269	9	8	1	2	1	—	841
	Cases	675	129	327	3	2799	2293	49	33	5	4	5	—	6322
	Incidence rate	19	5	32	—	26	32	7	4	4	1	8	—	22
1993	Outbreaks	200	79	21	2	103	259	6	9	2	2	—	—	683
	Cases	1097	336	124	7	1195	1620	54	24	11	7	—	—	4475
	Incidence rate	32	12	12	1	11	22	7	3	8	1	—	—	16
1994	Outbreaks	198	77	26	6	153	300	6	14	1	5	—	—	787
	Cases	1166	326	196	41	1910	2254	27	53	19	15	—	—	6040
	Incidence rate	32	12	19	4	17	31	4	6	14	3	—	—	21
1995	Outbreaks	204	11	22	4	139	267	9	22	—	1	3	1	684
	Cases	1705	107	125	13	2484	1900	66	90	—	4	55	3	6575
	Incidence rate	45	4	12	1	22	26	9	10	—	1	84	10	22

[a]Modified from Todd (1986, 1991), Todd & Chatman (1997, 1998) and Todd et al. (2000).
[b]Cases include numbers of cases in outbreaks and single cases.
[c]Number of cases per 100,000 population.

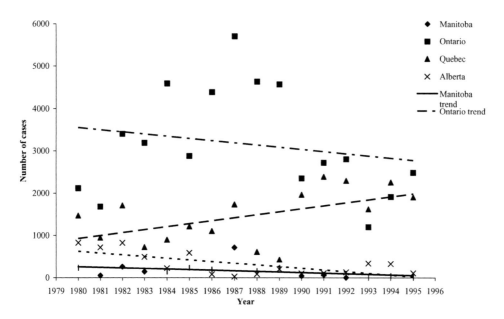

Figure 3.1. Foodborne disease cases in Manitoba, Ontario, Quebec, and Alberta, 1980–1995. Todd (1986, 1988, 1991), Todd & Harboway (1994), Todd & Chatman (1996, 1997, 1998) and Todd et al. (2000).

illness cases in the four major hog producing provinces during 1980 and 1995 are shown in Fig. 3.1. All except for Quebec show a downward trend in foodborne illness cases. Incidence rates of human *Salmonella*, *Campylobacter* and pathogenic *E. coli* infections by province and territory in 1995 (the most recent statistics) are presented in Table 3.14. *Campylobacter* is the most frequently reported enteric pathogen, followed by *Salmonella* and pathogenic *E. coli*. Most (91.3%) of the pathogenic *E. coli* isolates (n = 1,455) from human cases in 1995 were serovar O157 (H7 flagellar antigen was not specified) (Health Canada 1998a). It appears that cases of *Campylobacter* and *Salmonella* infection were fairly evenly distributed geographically, whereas incidence of pathogenic *E. coli* showed increased rates in central to western Canada and Prince Edward Island. This may be a result of increased livestock production in these regions. According to this data, Manitoba has the highest *E. coli* incidence rate per 100,000 population among all provinces and territories. However, variation in reported rates has to be carefully interpreted since there are huge differences in the within-province protocol for reporting among provinces. There are two sets of databases used in Tables 3.13 and 3.14, namely the National Notifiable Diseases system and the National Laboratory for Bacteriology and Enteric Pathogens. In some cases, the incidence rates of both databases are similar for some provinces but differ by a factor of 2 to 7 for other provinces. Therefore, oftentimes, the differences in number of cases between databases are the reflection of differences in extent of reporting among provinces.

Table 3.14. Rates of human cases (food and non-food sources) of salmonellosis,
***Campylobacter* and pathogenic *E. coli* infections by province, 1995.**[a]

Province/territories	*Salmonella* cases per 100,000 population (NNDS[b] data/ NLBEP[c] data)	*Campylobacter* cases per 100,000 population (NNDS data/ NLBEP data)	Pathogenic *E. coli* cases per 100,000 population (NNDS data/ NLBEP data)
Newfoundland	14.0/13.8	15.4/4.2	2.1/1.0
Prince Edward Island	17.0/18.5	36.9/36.2	5.9/8.1
Nova Scotia	16.1/15.5	24.9/8.1	1.9/0.9
New Brunswick	19.3/25.1	30.9/33.4	0.4/2.8
Quebec	**18.1/18.5**	**33.5/33.4**	**4.7/4.1**
Ontario	**25.8/29.9**	**57.4/57.1**	**5.2/5.8**
Manitoba	**17.3/19.6**	**22.6/8.7**	**14.0/10.9**
Saskatchewan	19.2/22.3	23.5/24.6	5.0/5.4
Alberta	17.1/23.8	33.0/7.1	4.5/5.2
British Columbia	23.8/28.8	58.1/15.4	5.2/3.7
North West Territories	37.8/—[d]	52.5/—	—/9.8
Yukon	32.8/—	13.1/—	—/—
National average	21.6 (n = 6,389)/ 24.7 (n = 7,307)	46.2 (n = 13,680)/ 34.8 (n = 10,320)	5.0 (n = 1,493)/ 4.9 (n = 1,455)

[a]Modified from Health Canada (1998a).
[b]National Notifiable Diseases system.
[c]National Laboratory for Bacteriology and Enteric Pathogens system.
[d]Not reported.

The ten most frequently isolated *Salmonella* serovars from human and non-human sources in Canada and Manitoba in 1995 are shown in Tables 3.15 and 3.16. Half of these serovars appear on both lists. For Canada, they are serovars Typhimurium, Heidelberg, Hadar, Thompson, and Agona. For

**Table 3.15. Top ten *Salmonella enterica*
serovars from human and non-human
sources reported in Canada, 1995.**[a]

Human sources	Non-human sources
Typhimurium	**Heidelberg**
Enteritidis	**Typhimurium**
Heidelberg	**Hadar**
Hadar	Anatum
Thompson	Kentucky
Agona	Muenster
Newport	Schwarzengrund
Typhi	Senftenberg
Infantis	**Thompson**
Saintpaul	**Agona**

[a]Modified from Health Canada (1998a).

Table 3.16. Top ten *Salmonella enterica* serovars from human and non-human sources reported in Manitoba, 1995.[a]

Human sources	Non-human sources
Typhimurium	Senftenberg
Hadar	Mbandaka
Enteritidis	Tennessee
Muenchen	**Typhimurium**
Heidelberg	Havana
Thompson	**Heidelberg**
Java	**Montevideo**
Istanbul	Cerro
Montevideo	**Hadar**
Agona	**Agona**

[a]Modified from Health Canada (1998b,c).

Manitoba, they are Typhimurium, Hadar, Heidelberg, Montevideo, and Agona. In Canada, *S.* Typhimurium is frequently isolated from cattle, swine, turkey, and chicken, whereas *S.* Enteritidis is isolated from chicken, swine, and cattle, in the order of decreasing isolation rates (Health Canada 1998a). *Salmonella* contamination of Canadian swine has varied over time. An early Canada-wide study by Lammerding et al. (1988) recovered *Salmonella* from 17.5% of pork carcasses at slaughter and *S.* Brandenburg, *S.* Derby, *S.* Infantis, and *S.* Typhimurium were the most frequently isolated serovars. A later study in Quebec swine herds (Letellier et al. 1999a) isolated *Salmonella* from 7.9% of all fecal samples (n = 1,923) and 70.7% of all farms sampled (n = 41). The predominant serovars were *S.* Derby and *S.* Typhimurium, which accounted for 37.1%, and 34.1% of all ten serovars identified (n = 132). A larger study by these researchers (Letellier et al. 1999b) on finishing pigs from Quebec, Ontario and Manitoba reported an overall *Salmonella* prevalence of 5.2%. In this study, the most frequently isolated serovars were *S.* Brandenburg (40.9%), *S.* Infantis (16.4), *S.* Derby (9.8), and *S.* Typhimurium (8.2). Later, Rheault & Quessy (1999) reported a *Salmonella* isolation rate of 21.3% in feces of finishing swine in four Quebec swine abattoirs. The authors believed that the unusually high prevalence of *Salmonella* in Quebec swine was related to the many clinical outbreaks of *S.* Typhimurium in the province at that time. However, detailed data were not presented to support this claim.

The predominant serotypes of *Y. enterocolitica* implicated in human illness in Europe are O:3 and O:9, while in the U.S. are O:8 and O:5,27. Earlier studies in Ontario, Quebec, P. E. I., and Saskatchewan in Canada have shown that O:3 was the most frequently isolated serotype from Canadian swine herds (Mafu et al. 1989; Kwaga et al. 1990; Schiemann & Fleming 1981). These studies found that between 11.9–19% of swine carcasses in Canada were contaminated by *Y. enterocolitica*. In particular, serotype O:3 was more prevalent in swine in eastern provinces whereas serotype O:5,27 was more common in western provinces (Schiemann & Fleming 1981). The incidence of the serotype in swine also correlated with the human incidence of the same serotype in respective provinces. Interestingly, although serotype O:8 was the most common human serotype in the western provinces, it was not isolated from swine in the survey. In one study, serotype O:3 and O:5 occurred equally frequent in retail pork products in Ontario (Schiemann 1980). A later study on finishing swine from Quebec, Ontario, and Manitoba reported an overall *Y. enterocolitica* prevalence of 20.9% (Letellier et al. 1999b). Of 275 isolates found in this study, 85.5% were serotype O:3, 9.1% were O:5, 3.3% were O:9, and 0.4% were O:8. A more recent study by Thibodeau et al. (2001) at a Quebec slaughterhouse found that 27% of swine carried *Y. enterocolitica* O:3 in their tonsils and/or intestinal tracts. These studies have confirmed the ubiquitous contamination by *Y. enterocolitica* O:3 of Canadian swine. In fact, researchers from Japan have reported the introduction of pathogenic *Y. enterocolitica* strains (serotypes O:3 and O:5,27) through pork imported from Canada, Denmark, Taiwan, and the United States (Fukushima et al. 1997).

As in the U.S., *Campylobacter* is also a leading cause of foodborne illness in Canada. *Campylobacter* contamination also appears to be common among Canadian swine. An early study in a Quebec abattoir isolated *Campylobacter* from 61.7% of all muscle and fecal samples (n = 400) (Mafu et al. 1989). *C. coli* and *C. jejuni* accounted for 97 and 2%, respectively. Ninety-nine percent of all fecal samples were positive for the presence of *C. coli*. This agrees with the later U.S. studies (Wesley et al. 1999; Wesley et al. 1998; McKean et al. 2000). In a national surveillance program across Canada during the years 1983 to 1986, thermophilic *Campylobacter* were isolated from 16.9% of pork collected from federally inspected slaughter establishments (Lammerding et al. 1988). *C. coli* were also the predominant thermophilic *Campylobacter* in pork in the survey.

Although Canada has the largest number of human VTEC infections, none of them has been associated with pigs and pork. Instead, direct or indirect exposure to cattle is an important potential source of infection. A few studies have tried to find a link between the livestock density and human VTEC infection. Michel et al. (1999) found that there was a positive association between cattle density and VTEC incidence of reported cases in Ontario between 1990 and 1995. In other words, the incidence of human VTEC infection was higher in rural areas than in urban areas of Ontario during this period. They also identified contact with cattle, consumption of contaminated well water or locally produced food as important risk factors for human VTEC infection, which they believed had been previously underestimated. According to a recent publication (Valcour et al. 2002), strong associations were found between human VTEC infection and the regional ratio of beef cattle numbers to human population as well as the application of manure to the surface of agricultural land by solid and liquid spreaders. This is the first report that has identified application of manure to land as a potential risk factor for human VTEC infection. Interestingly, the researchers found a negative association between swine density and human VTEC infection. They explained that this inverse relationship might simply be a result of a relative absence of cattle in areas where swine are intensively farmed. Swine have been known to harbour VTEC within their intestinal tract, and while they are not considered to be important reservoirs of *E. coli* O157:H7, the organism has been isolated from hogs (Gyles et al. 2002). The use of livestock density indicators by Valcour et al. (2002) may be a valuable approach for establishing a link between swine density and human infection caused by other important swine-borne pathogens such as *Salmonella* and *Yersinia*.

CONCLUSIONS

From this information, it seems that assessment of the potential impact of pork production on human health is complicated by several factors. Cases of pork-associated foodborne illness in some countries appear to be more influenced by the amount of pork consumption, the prevalence of pathogens in pigs, the hygiene at slaughterhouse level, or even cooking habits, rather than by pork production or the pig density. In all hog-producing countries studied in this report, pork production did not increase national pork consumption, instead it increased exports. This is supported by the observation that the major pork producing countries are also major pork exporters. The relationship between pork consumption and pork-associated foodborne illness is apparent in Danish and Dutch situations. Pork consumption is the highest among all meat consumed in both countries and pork is a major source of human enteric illness. All human *Yersinia* and most *S.* Typhimurium infections in these countries are attributed to pork consumption. The two pathogens are also highly prevalent in Danish and Dutch swine. Improved hygiene at the slaughterhouse level has led to a decline of pathogen prevalence in hog carcasses, and presumably human cases. This is best illustrated in the Dutch situation, where *Y. enterocolitica* contamination of pigs is high, but carcass contamination at slaughter and human *Yersinia* cases are low. It is, however, a different scenario in Taiwan. Although pork consumption is also the highest in Taiwan, pork does not present a major source of human illness. Instead, seafood is the major vehicle of foodborne illness in this country. Because prevalence data of pathogens in pigs is not available for this country, evaluation cannot be made regarding its effect on human illness. Nonetheless, the high pig density and pork production in this country have not increased known domestic cases of pork-associated human illness. In Canada and the U.S., pork consumption ranks third after poultry and beef, and is relatively lower than all other pork producing countries considered in this report. Cases of human illness attributed to pork fluctuated in the U.S., despite the high pathogen prevalence in pigs and retail pork. Increased pork production in the U.S. did not seem to increase cases of pork-associated human illness. Lower pork consumption and thorough cooking of pork may have contributed to the lower incidence rates in the U.S. compared to other countries. A similar situation was observed in Canada, except that cases of human enteric illness from pork showed a decrease between 1975 and 1995. Thus, an increase in national pork production during this period has not resulted in additional cases of pork-associated human illness. However, a considerable increase in Canadian pork production occurred after 1996 (Table 2.2), but no foodborne illness data are available to assess the effect of this increase in pork production. In addition, the Canadian system of reporting foodborne illness varies from province to province, and this has most probably resulted in serious underreporting of foodborne diseases. This hypothesis is supported by the variable differences observed between the two national databases for reportable diseases. Lack of provincial information on pork-associated human cases may also have underestimated the impact of pork production in major pork producing provinces such as Quebec, Ontario and Manitoba. Given the currently available data, pork is not considered a major source of foodborne infection in Canada. An accurate assessment of recent increases of pork production upon human enteric illness must await collection and reporting of epidemiological data.

REFERENCES

Anonymous. 1997. The hog and pork industries of Denmark and the Netherlands: a competitiveness analysis [Online]. URL: http://www.agr.gc.ca/policy/epad/english/pubs/adhoc/pork/porktoc.htm (Accessed: May 20, 2002).

Anonymous. 1998. Annual report on zoonoses in Denmark 1997. Ministry of Food, Agriculture and Fisheries, Copenhagen, Denmark.

Anonymous. 1999. Annual report on zoonoses in Denmark 1998. Ministry of Food, Agriculture and Fisheries, Copenhagen, Denmark.

Anonymous. 2000. Annual report on zoonoses in Denmark 1999. Ministry of Food, Agriculture and Fisheries, Copenhagen, Denmark.

Anonymous. 2001. Annual report on zoonoses in Denmark 2000. Ministry of Food, Agriculture and Fisheries, Copenhagen, Denmark.

Anonymous. 2002. Annual report on zoonoses in Denmark 2001. Ministry of Food, Agriculture and Fisheries, Copenhagen, Denmark.

Berends, B. R., H. A. P. Urlings, J. M. A. Snijders, and F. van Knapen. 1996. Identification and quantification of risk factors in animal management and transport regarding *Salmonella* spp. in pigs. Int. J. Food Microbiol. 30:37–53.

Berends, B. R., F. van Knapen, J. M. A. Snijders, and D. A. A. Mossel. 1997. Identification and quantification of risk factors regarding *Salmonella* spp. on pork carcasses. Int. J. Food Microbiol. 36:199–206.

Berends, B. R., F. van Knapen, D. A. A. Mossel, S. A. Burt, and J. M. A. Snijders. 1998. Impact on human health of *Salmonella* spp. on pork in the Netherlands and the anticipated effects of some currently proposed control strategies. Int. J. Food Microbiol. 44:219–229.

Center for Science in the Public Interest (CSPI). 2001. Outbreak alert—closing the gaps in our federal food-safety net [Online]. URL: http://www.cspinet.org/reports/oa_2001.pdf (Accessed: June 4, 2002).

The Centers for Disease Control and Prevention (CDC). 2001. Preliminary FoodNet data on the incidence of foodborne illnesses—selected sites, United States, 2000. Morb. Mortal. Wkly. Rep. 50:241–246.

CDC. 2002. Preliminary FoodNet data on the incidence of foodborne illnesses—selected sites, United States, 2001. Morb. Mortal. Wkly. Rep. 51:325–329.

Chiou, A., L. H. Chen, and S. H. Chen. 1991. Foodborne illness in Taiwan, 1981–1989. Food Aust. 43:70–71.

Chiou, C. S., S. Y. Hsu, S. I. Chiu, T. K. Wang, and C. S. Chao. 2000. *Vibrio parahaemolyticus* serovar O3:K6 as cause of unusually high incidence of food-borne disease outbreaks in Taiwan from 1996 to 1999. J. Clin. Microbiol. 38: 4621–4625.

Chiu, C. H., T. L. Wu, L. H. Su, C. Chu, J. H. Chia, A. J. Kuo, M. S. Chien, and T. Y. Lin. 2002. The emergence in Taiwan of fluoroquinolone resistance in *Salmonella enterica* serotype Choleraesuis. N. Engl. J. Med. 346:413–419.

DANMAP. 2000. Consumption of antimicrobial agents and occurrence of antimicrobial resistance in bacteria from food animals, foods and humans in Denmark [Online]. URL: http://www.svs.dk/dk/Publikationer/Danmap/Danmap%202000.pdf (Accessed: May 17, 2002).

de Boer, E., W. M. Seldam, and J. Oosterom. 1986. Characterization of *Yersinia enterocolitica* and related species isolated from foods and porcine tonsils in the Netherlands. Int. J. Food Microbiol. 3:217–224.

de Boer, E., and J. F. M. Nouws. 1991. Slaughter pigs and pork as a source of human pathogenic *Yersinia enterocolitica*. Int. J. Food Microbiol. 12:375–378.

Duffy, E. A., K. E. Belk, J. N. Sofos, G. R. Bellinger, A. Pape, and G. C. Smith. 2001. Extent of microbial contamination in United States pork retail products. J. Food Protect. 64: 172–178.

Fedorka-Cray, P., J. D. McKean, G. W. Beran. 1997. Prevalence of *Salmonella* in swine and pork: a farm to consumer study [Online]. Iowa State University Swine Research Report. URL: http://www.extension.iastate.edu/Pages/ansci/swinereports/asl-1507.pdf (Accessed: June 3, 2002).

Fedorka-Cray, P., E. Bush, L. A. Thomas, J. T. Gray, J. McKean, and D. L. Harris. 1996. *Salmonella* infection in herds of swine [Online]. Iowa State University Swine Research Report. URL: http://www.extension.iastate.edu/Pages/ansci/swinereports/asl-1412.pdf (Accessed: June 3, 2002).

Food Safety and Inspection Service (FSIS). 1996. Nationwide pork microbiological baseline data collection program: market hogs [Online]. URL: http://www.fsis.usda.gov/OPHS/baseline/markhog1.pdf and http://www.fsis.usda.gov/OPHS/baseline/markhog2.pdf (Accessed: April 11, 2002).

FSIS. 2002. Progress report on *Salmonella* testing of raw meat and poultry products, 1998–2001 [Online]. URL: http://www.fsis.usda.gov/ophs/haccp/salm4year.htm (Accessed: May 1, 2002).

Fukushima, H., K. Hoshina, H. Itogawa, and M. Gomyoda. 1997. Introduction into Japan of pathogenic *Yersinia* through imported pork, beef, and fowl. Int. J. Food Microbiol. 35:205–212.

Funk, J. A., H. F. Troutt, R. E. Isaacson, and C. P. Fossler. 1998. Prevalence of pathogenic *Yersinia enterocolitica* in groups of swine at slaughter. J. Food Protect. 61:677–682.

Gyles, C. L., R. Friendship, K. Ziebell, S. Johnson, I. Yong, and R. Amezcua. 2002. *Escherichia coli* O157:H7 in pigs. Expert Committee on Meat and Poultry Products, Canada Committee on Food, Can. Agri-Food Res Council, September 16–17, Guelph ON.

Hald, T., and J. S. Andersen. 2001. Trends and seasonal variation in the occurrence of *Salmonella* in pigs, pork and humans in Denmark, 1995–2000. Berl. Munch. Tierarztl. Wschr. 114:346–349.

Hald, T., and H. C. Wegener. 1999. Quantitative assessment of the sources of human salmonellosis attributable to pork [Online]. Proceedings, 3rd international symposium on the epidemiology and control of Salmonella in pork, Washington D. C., August 5–7, 1999. URL: http://www.isecsp99.org (Accessed: April 17, 2002).

Health Canada. 1998a. Canadian integrated surveillance report for 1995 on *Salmonella, Campylobacter*, and pathogenic *E. coli*. Can. Commun. Dis. Rep. Suppl. 24S5.

Health Canada. 1998b. Most frequently isolated *Salmonella* serotypes from various provinces 1995 (human) [Online]. URL: http://www.hc-sc.gc.ca/hpb/lcdc/bmb/epiic95/95_95_o_e.html (Accessed: June 10, 2002).

Health Canada. 1998c. Most frequently isolated *Salmonella* serotypes from various provinces 1995 (non-human) [Online]. URL: http://www.hcsc.gc.ca/hpb/lcdc/bmb/epiic95/95_95_t_e.html (Accessed: June 10, 2002).

Heuvelink, A. E., J. T. M. Zwartkruis-Nahuis, F. L. A. M. van den Biggelaar, W. J. van Leeuwen, and E. de Boer. 1999a. Isolation and characterization of verocytotoxin-producing *Escherichia coli* O157 from slaughter pigs and poultry. Int. J. Food Microbiol. 52:67–75.

Heuvelink, A. E., J. T. M. Zwartkruis-Nahuis, R. R. Beumer, and E. de Boer. 1999b. Occurrence and survival of verocytotoxin-producing *Escherichia coli* O157 in meats obtained from retail outlets in the Netherlands. J. Food Protect. 62:1115–1122.

Hurd, H. S., J. D. McKean, I. V. Wesley, and L. A. Karriker. 2001. The effect of lairage on *Salmonella* isolation from market swine. J. Food Protect. 64:939–944.

Iowa Department of Agriculture & Land Stewardship (IDALS). 2001. Iowa agriculture quick facts [Online]. URL: http://www.agriculture.state.ia.us/quickfacts.htm (Accessed: June 7, 2002).

Iowa Department of Public Health (IDPH). 2001. Reportable diseases by decade [Online]. URL: http://www.idph.state.ia.us/pa/ic/decades.pdf (Accessed: June 4, 2002).

Iowa State University (ISU). 1997. ISU food safety consortium program accomplishments [Online]. URL: http://www.foodsafety.iastate.edu/annrpts.html (Accessed: June 6, 2002).

Isaacson, R. E. 2000. Stealth *Salmonella*: problems for detecting *Salmonella* positive animals [Online]. A workshop on epidemiologic methods and approaches for food safety, October 18–19, 2000, Birmingham, Alabama. URL: http://vbms.unl.edu/wills/epiconf/text/4%20isaacson.pdf (Accessed: June 18, 2002).

Kwaga, J., J. O. Iversen, and J. R. Saunders. 1990. Comparison of two enrichment protocols for the detection of *Yersinia* in slaughtered pigs and pork products. J. Food Protect. 53:1047–1049.

Lammerding, A. M., M. M. Garcia, E. D. Mann, Y. Robinson, W. J. Dorward, R. B. Truscott, and F. Tittiger. 1988. Prevalence of *Salmonella* and thermophilic *Campylobacter* in fresh pork, beef, veal and poultry in Canada. J. Food Protect. 51:47–52.

Lee, L. A., J. Taylor, G. P. Carter, B. Quinn, J. J. Farmer III, and R. V. Tauxe. 1991. *Yersinia enterocolitica* O:3: an emerging cause of pediatric gastroenteritis in the United States. J. Infect. Dis. 163:660–663.

Letellier, A., S. Messier, J. Pare, J. Menard, and S. Quessy. 1999a. Distribution of *Salmonella* in swine herds in Quebec. Vet. Microbiol. 67:299–306.

Letellier, A., S. Messier, and S. Quessy. 1999b. Prevalence of *Salmonella* spp. and *Yersinia enterocolitica* in finishing swine at Canadian abattoirs. J. Food Protect. 62:22–25.

Lu, P. L., P. R. Hsueh, C. C. Hung, S. C. Chang, K. T. Luh, and C. Y. Lee. 2000. Bacteremia due to *Campylobacter* species: high rate of resistance to macrolide and quinolone antibiotics. J. Formosa. Med. Assoc. 99:612–617.

Mafu, A. A., R. Higgins, M. Nadeau, and G. Cousineau. 1989. The incidence of *Salmonella, Campylobacter*, and *Yersinia enterocolitica* in swine carcasses and the slaughterhouse environment. J. Food Protect. 52:642–645.

McKean, J. D., G. W. Beran, T. Proescholdt, P. Davies, P. Turkson, J. Kliebenstein, L. J. Hoffman, and J. S. Dickson. 2000. The prevalence of food-borne pathogenic organisms in swine and pork: a pilot survey and demonstration project from production farm to dressed carcasses [Online]. Iowa State University Swine Research Report. URL: http://www.extension.iastate.edu/ipic/reports/00swinereports/asl-693a.pdf (Accessed: June 3, 2002).

Michel, P., J. B. Wilson, S. W. Martin, R. C. Clarke, S. A. McEwen, and C. L. Gyles. 1999. Temporal and geographical distributions of reported cases of *Escherichia coli* O157:H7 infection in Ontario. Epidemiol. Infect. 122:193–200.

Moyer, N. P. 1997a. Top 10 *Salmonella* serotypes in Iowa—1993—1995 [Online]. URL: http://www.uhl.uiowa.edu/publications/hotline/1997_11/salmonella.html (Accessed: June 3, 2002).

Moyer, N. P. 1997b. *Salmonella* surveillance data—1995 [Online]. URL: http://www.uhl.uiowa.edu/publications/hotline/1997_12/salmonella.html (Accessed: June 3, 2002).

Nielsen, B., and H. C. Wegener. 1997. Public health and pork and pork products: regional perspectives of Denmark. Rev. Sci. Technol. Off. Int. Epiz. 16:513–524.

Pan, T. M., T. K. Wang, C. L. Lee, S. W. Chien, and C. B. Horng. 1997. Food-borne disease outbreaks due to bacteria in Taiwan, 1986 to 1995. J. Clin. Microbiol. 35: 1260–1262.

Potter, M., and IFT expert report panelists. 2002. Emerging microbiological food safety issues: implications for control in the 21st century [Online]. IFT expert report. URL: http://www.ift.org (Accessed: February 25, 2002).

Rheault, N., and S. Quessy. 1999. Sampling of environment and carcasses for the detection of *Salmonella* in swine abattoirs [Online]. Proceedings of the 3rd international symposium on the epidemiology and control of *Salmonella* in pork, Washington, D.C., August 5–7, 1999. URL: http://www.isecsp99.org (Accessed: May 27, 2002).

Saskatchewan Agriculture and Food (SAF). 2001. Opportunities in Saskatchewan [Online]. URL: http://www.agr.gov.sk.ca/DOCS/livestock/pork/production_information/Porkopportunities2001.pdf (Accessed: May 17, 2002).

Schiemann, D. A. 1980. Isolation of toxigenic *Yersinia enterocolitica* from retail pork products. J. Food Protect. 43:360–365.

Schiemann, D. A., and C. A. Fleming. 1981. *Yersinia enterocolitica* isolated from throats of swine in eastern and western Canada. Can. J. Microbiol. 27:1326–1333.

Statistics Canada. 2000. Food consumption in Canada, Part 1 [Online]. URL: http://www.statcan.ca/english/IPS/Data/32-229-XIB.htm (Accessed: October 28, 2002).

Statistics Canada. 2001. Food consumption in Canada, Part II [Online]. URL: http://www.statcan.ca/english/IPS/Data/32-230-XIB.htm (Accessed: October 28, 2002).

Statistics Netherlands. 2001. Livestock, cattle and pigs [Online]. URL: http://www.cbs.nl/en/figures/keyfigures/llb0005t.htm (Accessed: May 22, 2002).

Thibodeau, V., E. H. Frost, and S. Quessy. 2001. Development of an ELISA procedure to detect swine carriers of pathogenic *Yersinia enterocolitica*. (Abs.) Vet. Microbiol. 82: 249–259.

Todd, E. C. D. 1986. Foodborne and waterborne disease in Canada, annual summaries 1980, 1981 and 1982. Health Protection Branch, Health and Welfare Canada, Polyscience Publications Inc., Morin Heights, Quebec.

Todd, E. C. D. 1991. Foodborne and waterborne disease in Canada, annual summaries 1985 and 1986. Health Protection Branch, Health and Welfare Canada. Polyscience Publications Inc., Morin Heights, Quebec.

Todd, E. C. D., and P. Chatman. 1997. Foodborne and waterborne disease in Canada, annual summaries 1990 and 1991. Health Protection Branch, Health Canada. Polyscience Publications Inc., Morin Heights, Quebec.

Todd, E. C. D., and P. Chatman. 1998. Foodborne and waterborne disease in Canada, annual summaries 1992 and 1993. Health Protection Branch, Health Canada. Polyscience Publications Inc., Laval, Quebec.

Todd, E. C. D.; P. Chatman, and V. Rodrigues. 2000. Annual summaries of foodborne and waterborne disease in Canada, 1994 and 1995, Health Products and Food Branch, Health Canada. Polyscience Publications Inc., Laval, Quebec.

United States Animal Health Association (USAHA). 2001. Report of the committee on *Salmonella* [Online]. 2001 Committee Reports. URL: http://www.usaha.org/reports/reports01/r01sal.html (Accessed: June 3, 2002).

United States Department of Agriculture (USDA). 1997. Pork summary-selected countries [Online]. Foreign Agricultural Service. URL: http://www.fas.usda.gov/dlp2/circular/1997/97-03/porksumm.htm (Accessed: April 3, 2002).

USDA. 1998. Per capita pork consumption—selected countries [Online]. Foreign Agricultural Service. URL: http://www.fas.usda.gov/dlp2/circular/1998/98-03LP/Tables/table21.pdf (Accessed: May 17, 2002).

USDA. 2000. Pork summary-selected countries [Online]. Foreign Agricultural Service. URL: http://www.fas.usda.gov/dlp/circular/2000/00-03lp/porkpr.pdf (Accessed: April 3, 2002).

USDA. 2001a. Pork summary—selected countries [Online]. Foreign Agricultural Service. URL: http://www.fas.usda.gov/circular/2001/01–03lp/livestock.html (Accessed: April 3, 2002).

USDA. 2001b. Per capita pork consumption—selected countries [Online]. Foreign Agricultural Service. URL: http://www.fas.usda.gov/circular/2001/01-03lp/porkpcc.pdf (Accessed: April 3, 2002).

USDA. 2001c. Per capita beef and veal consumption—selected countries [Online]. Foreign Agricultural Service. URL: http://www.fas.usda.gov/circular/2001/01-03lp/beefpcc.pdf (Accessed: May 17, 2002).

USDA. 2001d. Ecology and epidemiology of *Salmonella* and other foodborne pathogens in livestock. National Animal Disease Center. Pre-Harvest Food Safety & Enteric Disease Research Unit [Online]. URL: http://www.nadc.ars.usda.gov/research/fs/epidemiology/ (Accessed: April 1, 2002).

USDA. 2002a. Pork summary-selected countries [Online]. Foreign Agricultural Service. URL: http://www.fas.usda.gov/dlp/circular/2002/02-03LP/pk_sum.pdf (Accessed: April 3, 2002).

USDA. 2002b. Livestock slaughter 2001 summary [Online]. National Agricultural Statistics Service. URL: http://usda.mannlib.cornell.edu/reports/nassr/livestock/pls-bban/lsan0302.pdf (Accessed: June 4, 2002).

Valcour, J. E., P. Michel, S. A. McEwen, and J. B. Wilson. 2002. Associations between indicators of livestock farming intensity and incidence of human shiga toxin-producing *Escherichia coli* infection. Emerg. Infect. Dis. 8:252–257.

van der Wolf, P. J., J. H. Bongers, A. R. W. Elbers, F. M. M. C. Franssen, W. A. Hunneman, A. C. A. van Exsel, and M. J. M. Tielen. 1999. *Salmonella* infections in finishing pigs in the Netherlands: bacteriological herd prevalence, serogroup and antibiotic resistance of isolates and risk factors for infection. Vet. Microbiol. 67:263–275.

van der Wolf, P. J., A. R. W. Elbers, H. M. J. F. van der Heijden, F. W. van Schie, W. A. Hunneman, and M. J. M. Tielen. 2001. *Salmonella* seroprevalence at the population and herd level in pigs in the Netherlands. Vet. Microbiol. 80:171–184.

Wayt, B. 2001. Comment on an issue of INSIDER magazine "Scotland's National Business Magazine" [Online]. URL: http://www.sovereignty.org.uk/features/footmouth/taiwan.html (Accessed: May 17, 2002).

Wesley, I. V., K. M. Harmon, A. Green, E. Bush, and S. Wells. 1999. Distribution of *Campylobacter* and *Arcobacter* in livestock [Online]. Iowa State University Swine Research Report. URL: http://www.extension.iastate.edu/ipic/reports/99swinereports/asl-1707.pdf (Accessed: June 3, 2002).

Wesley, I. V., J. McKean, P. Turkson, P. Davies, S. Johnson, T. Proescholdt, and G. Beran. 1998. *Campylobacter* spp. and *Yersinia enterocolitica* in growing pigs in Iowa and North Carolina: a pilot study [Online]. Iowa State University Swine Research Report. URL: http://www.extension.iastate.edu/Pages/ansci/swinereports/asl-1604.pdf (Accessed: June 3, 2002).

Wilson, T. M., and C. Tuszynski. 1997. Foot-and-mouth disease in Taiwan—1997 overview [Online]. URL: http://www.usaha.org/reports/taiwanfmd.html (Accessed: May 17, 2002).

World Health Organization (WHO). 1999. WHO surveillance programme for control of foodborne infections and intoxications in Europe 7th report [Online]. URL: http://www.bgvv.de/publikationen/who/7threport/7threp_fr.htm (Accessed: May 15, 2002).

World Health Organization (WHO). 2002. WHO surveillance programme for control of foodborne infections and intoxications in Europe 8th report (in preparation).

Zhao, C., B. Ge, J. de Villena, R. Sudler, E. Yeh, S. Zhao, D. G. White, D. Wagner, and J. Meng. 2001. Prevalence of *Campylobacter spp.*, *Escherichia coli*, and *Salmonella* serovars in retail chicken, turkey, pork, and beef from the Greater Washington, D.C., area. Appl. Environ. Microbiol. 67:5431–5436.

CHAPTER 4

Reliability of Detection Methods for Pathogens and Availability of Molecular Methods

Detection methods for pathogens from animals or the environment are generally characterized by their microbiological sensitivity, as well as their epidemiologic sensitivity and specificity. Isaacson (2000) defined the epidemiologic sensitivity of a detection method as a combination of how few organisms can be detected by an assay (microbiological sensitivity), how well the assay performs using 'real' samples obtained from animals, whether a sample collected from a 'positive animal' is positive at the time of sampling, and whether a proper sample has been collected. Specificity of a detection assay was defined as how well the assay discriminates between the desired target organism and other related but incorrect targets. The reliability of any detection method is affected by issues concerning its sensitivity and specificity.

DETECTION OF BACTERIA

Detection of *Salmonella* in swine is a good example for use when examining sample criteria as it has merited a lot of attention. Culturing fecal samples has traditionally been used to detect *Salmonella*-infected animals on farm and still remains the gold standard today. However, culture methods will not identify all infected animals as the bacteria can be shed intermittently in the feces. Animals with clinical salmonellosis are easier to detect because they exhibit visible signs and usually shed high numbers of the pathogen in their feces, have large numbers in their gastrointestinal tracts, and sometimes in their blood and peripheral tissues, making culturing of *Salmonella* reasonably easy (Isaacson 2000).

However, asymptomatic or subclinical *Salmonella* infection in pigs is a more important public health problem associated with pork. Clinically healthy animals present a real challenge in accurately assessing their *Salmonella* status. Detection of the pathogen in carriers by direct culturing is complicated by being both time consuming and irreproducible (Isaacson 2000).

Davies et al. (2000) compared the ability of three culture methods for the detection of *Salmonella* in swine. They also compared the degree of agreement between two different laboratories using the same samples. Their results showed that increasing the size of the sample (from 1 to 10 g feces) dramatically increased sensitivity of the test (from 6 to 20.4% positive samples). In addition to the finding that all three assays have different sensitivities, the researchers also found that within a single protocol, there was a lack of reproducibility from one laboratory to another. Besides these, there are problems of reproducibility on repeated sampling. In a study by Isaacson et al. (1999), culture results of lymph node samples collected from pigs at slaughter showed that 67% of the herds were positive for *Salmonella*. However, re-testing the same herds in the following year they found that some of those designated as *Salmonella* positive earlier were subsequently negative and some of the previously negative herds were presently positive (unpublished result from Bahnson et al. cited by Isaacson 2000). It thus seems that assigning a herd as either *Salmonella* positive or negative is only an indicator of its *Salmonella* status at a specific time and cannot be used to predict its *Salmonella* status at another time. Further, sometimes a negative result for *Salmonella* in pigs or herds may not actually mean that the pigs or herds are free from *Salmonella*, rather it may mean that the pigs or herds are *Salmonella* carriers and are not actively shedding the organism in their feces.

One problem associated with the sensitivity of any detection method is what should be sampled (Isaacson 2000). In most studies reported, feces, mesenteric lymph nodes, and intestinal contents were sampled to determine whether pigs were carriers of *Salmonella*. The choice of sample should depend on the specific question being addressed. For example, the use of lymph nodes has an advantage of being a 'clean' sample (contains relatively fewer other bacteria compared to feces), thus it is reasonably easy to culture (Isaacson 2000). However lymph node samples can only be collected at slaughter, and is not appropriate for determination of on farm prevalence. If the question is about prevalence at slaughter, then it is an appropriate choice. On the other hand, fecal samples can be collected on farm anytime but sample results only represent what is being shed at that particular time. Feces are thus considered a poor indicator of pathogen prevalence in pigs on the farm. The use of intestinal contents has the advantage of representing a real source of contamination but as with lymph nodes, samples can only be collected at slaughter and this approach could not be used on farm. Therefore, it appears there is no ideal sample that can be used for an accurate determination of *Salmonella* prevalence on farm.

Other factors that can affect the sensitivity of culture methods include fecal sample storage, type of enrichment broths or plating media, and the viable but nonculturable (VBNC) state of bacteria. Davies et al. (1999) reported that immediate processing of samples yielded the best recovery of *Salmonella*, while freezing of fecal samples resulted in significant reduction of detection. Storage of fecal samples at 4°C for several days did not significantly reduce detection. The authors also found (using the same sample weight), that primary enrichments in tetrathionate and Hajna GN broths and secondary enrichment in Rappaport Vassiliadis (RV) broth yielded significantly better detection of *Salmonella* than when pre-enrichment was in buffered peptone water followed by selective enrichment in RV broth. Waltman & Mallinson (1995) found a large variation in isolation protocols from veterinary laboratories across the U.S. for culturing poultry tissue and environmental samples for *Salmonella*. They found that there were 17 different selective enrichment media or combinations of enrichment media, and 14 different plating media being used. Variations were also found in the incubation period and temperature used for selective enrichments. The authors argued that it is important for diagnostic laboratories to adopt standardized protocols for isolating *Salmonella*. Similarly, Wesley et al. (1997) compared the sensitivity of three enrichments for *Yersinia enterocolitica*, namely irgasan, ticarsillin, plus potassium chlorate (ITC), modified trypticase soy broth (MTSB), and phosphate buffered saline (PBS). Their results showed that ITC appeared to be the most sensitive among the three enrichments. These types of methodological factors may have contributed to the variable results of pathogen prevalence published for swine and probably other animal species as well. Finally one other major concern with culture methods is the VBNC state of stressed or injured bacteria which will yield an underestimate of the results. Staining and microscopy methods are available for their detection but are rarely used by researchers (Crook et al. 1998).

Most techniques for the detection of *Campylobacter* spp. follow a conventional approach which includes pre-enrichment and enrichment followed by growth on selective media and biochemical confirmation. Selective media make use of a cocktail of antibiotics in a rich basal medium and rely on the ability of the organisms to grow at 42°C (Uyttendaele et al. 1997). One study (Scotter et al. 1993) compared three culture methods (two of which were proposed by the International Standards Organization) for the detection of thermotolerant campylobacters. The results showed that one of the two methods proposed by ISO was significantly more sensitive than the other two. The major differences used in this method were a pre-enrichment of samples with a gradual addition of antibiotics to suppress competing organisms and use of a non-selective blood agar isolation medium in combination with a membrane filtration technique. However, it still produced 18% false-negative (compared to 48–54% with the other two) and 8% false-positive results. Other researchers (Stern & Line 1992) who compared different enrichment methods also confirmed the need for enrichments to detect *Campylobacter*. In addition, they suggested an extra dilution of enrichment prior to plating on selective media so that the antimicrobial agents could perform more effectively against competing flora and so that *Campylobacter* spp. could compete better for the low level of oxygen in the test environment. Beuchat (1985) also noted that carbon dioxide enhanced recovery of *C. jejuni* by inhibiting growth of aerobes that could negatively affect viability of the target organism. Inability to compete well in the presence of other microorganisms as well as its microaerophilic nature has made isolation and detection of *Campylobacter* additionally challenging with culture techniques. The fastidious growth requirement of *Campylobacter* may have contributed to its poor recovery from environmental samples, which is not a natural habitat of most *Campylobacter* species. *Campylobacter* is believed to be a poor survivor outside its natural habitats, the intestines of birds and other warm-blooded animals. However, its poor survivability in the environment may have been a direct result of the poor recovery due to the inadequate detection methods used. In order to survive in hostile environmental conditions, the organism is known to enter the VBNC state, changing its morphology from a culturable spiral form to non-culturable coccoid forms. VBNC cells cannot be detected by traditional culture methods although they are considered potentially pathogenic when conditions become favourable. VBNC *Campylobacter* in water have been successfully detected by immunofluorescent-antibody and—rRNA hybrid staining microscopy (Buswell et al. 1998). With the advances in new detection techniques such as the immunochemical and molecular methods, the survival of *Campylobacter* spp. in the environment will be better characterized.

The standard culture method for the detection of *E. coli* O157:H7 includes growth of non-sorbitol-fermenting *E. coli* colonies on sorbitol MacConkey agar culture (SMAC), followed by serological confirmation with O157- and H7- specific antisera (Louie et al. 1998). The culture method is insensitive especially for the detection of small numbers of *E. coli* O157:H7, and is unable to detect non-O157 verotoxin-producing *E. coli* (Novicki et al. 2000). Therefore, the inability to simultaneously identify more than one species of bacteria and differentiate between species is another common disadvantage of conventional culture methods. Culture techniques are often insensitive, yield variable results, and can be non-specific, expensive, and time consuming. Sampling methods used also varied greatly among reported epidemiologic studies of pathogens, limiting the ability to compare results. One thing is apparent, that if a culture method is chosen, there is at least a need to adopt standardized sampling protocols as a first step in utilizing results to best advantage.

Serology has been developed as a more sensitive approach for detecting *Salmonella* infection status in animals. Antibodies to *Salmonella* last approximately three months and are present if the animal has been exposed to the bacterium (Rajic & Keenliside 2001). Thus, enzyme-linked immunosorbent assay (ELISA) based detection of *Salmonella* has the advantage of easily allowing analysis of large numbers of samples and is less expensive compared to culture methods (Isaacson 2000). Serologic testing is extensively used in Denmark as a basis of remediation practices on farm to reduce the presence of *Salmonella* in swine. This method is also useful in measuring the effectiveness of on farm intervention practices with herds in order to reduce the presence of *Salmonella*. In addition to its use in swine, serologic surveillance of *Salmonella* infection has also been applied to cattle and poultry. In a Dutch study (van Zijderveld et al. 1992), an ELISA assay was successfully developed that detects *S.* Enteritidis infections in experimentally infected chickens. The assay was reported to have 100% sensitivity and specificity. The use of ELISA for the detection of *S*. Dublin in carrier cattle was demonstrated by Spier et al. (1990), but its sensitivity and specificity were not determined.

Serological methods allow for detection of more than one serotype of a bacterial species. A Danish enzyme-linked immunosorbent assay (ELISA) was formulated specifically to detect pigs infected with *S*. Typhimurium and *S*. Infantis (Nielsen et al. 1995). While conventional culture methods detect only *E. coli* O157:H7, immunoassays have been documented for the detection of all enterohemorrhagic *Escherichia coli* (EHEC) serotypes (Kehl et al. 1997). Recently, a two-step method that combines ELISA toxin testing and the culture method was found able to detect and isolate all EHEC serotypes including serotype O157:H7 (Novicki et al. 2000). While conventional sorbitol MacConkey (SMAC) agar reliably detects only serotype O157:H7, this combined method utilizes a chromogenic selective-differential medium (Rainbow agar) for the isolation of *E. coli* O157:H7 together with an ELISA that detects the shiga-like toxins stx 1 and 2 (i.e. all EHEC serotypes).

A number of monoclonal antibodies have been used in an ELISA for rapid detection of thermophilic campylobacters, i.e. *C. jejuni*, and *C. lari* (Lu et al. 1997). However the detection limit of these antibodies fell in the range of 10^5–10^7 CFU/ml. Since only several hundred *Campylobacter* ($\leq$200 CFU/ml) are sufficient to cause infection in humans, more sensitive methods are needed to detect low numbers of these organisms in contaminated samples. More recently, Wang et al. (2000) demonstrated the use of a commercial polyclonal antibody in a dot-blot format enzyme immunoassay in combination with hydrophobic grid membrane filters (HGMF), which successfully detected all nine *Campylobacter* spp. in inoculated food samples. Although it still yielded 1.5% of false-positive results, it was much more specific than culture methods which generated about 50% false-positive results (see above). In the same study (Wang et al. 2000), approximately 130 non-*Campylobacter* strains were found able to grow on modified *Campylobacter* agar with charcoal and deoxycholate used as selective agents for *Campylobacter*. An ELISA method for detection of all thermotolerant *Campylobacter* spp. has been automated and is available commercially. It has been proven to be a rapid, reproducible and reliable method (Lilja & Hanninen 2001). Another immunochemical method involving latex-based agglutination has been made available commercially and is able to detect *C. jejuni* in water (Sutcliffe et al. 1991). However, because the technique exploits the use of antibody, antigens from dead *C. jejuni* cells can also be detected. Therefore, although the test is species specific, it is incapable of determining the viability of the cells present.

Using culture and serologic techniques, Baum et al. (1996) found a direct correlation between the presence of *Salmonella* in ileocecal lymph nodes of pigs (culture method) and the titer of antibody to *Salmonella* in meat juice (serologic), but there was no correlation between antibody titer in serum collected prior to slaughter (serologic) and the presence of *Salmonella* in the feces (culture). Therefore, the authors suggested that the ELISA method could be used to monitor the *Salmonella* status of farms by examining meat juice samples of pigs collected at slaughter. However, as with most detection methods, serologic detection also has its limitations. For *Salmonella*, the detection of anti-*Salmonella* antibodies from swine is only an indicator of previous exposure to *Salmonella* and does not necessarily mean that the pig is a carrier. As for detection of *E. coli* O157, the ELISA method may give false positive results due to *Citrobacter freundii*, *Escherichia hermanni*, and *Salmonella urbana* (Park et al. 1996). It was suggested that these non-*E. coli* strains may share O157 antigens. Given the uncertainties of serologic detection, it is important that new methods for detection of pathogens be pursued. New techniques that show promise are molecular methods such as polymerase chain reaction (PCR), DNA hybridization (gene probes), or biological sensors (biosensors).

PCR can be used in either single or multiple gene target (multiplex) formats. Both formats have been used for detection of various EHEC-associated gene sequences including the stx1 and 2 genes that are responsible for shiga-like toxin production. Louie et al. (1998) documented the use of a multiplex PCR for the detection of *E. coli* O157:H7 and verotoxin-producing non-O157 *E. coli* serotypes such as O26:H11 that are not readily detected by conventional culture methods. Boyapalle et al. (1999) compared multiplex and 5′ nuclease PCR assays with a bacteriological culture method for the rapid detection of pathogenic *Y. enterocolitica* in market weight hogs and pork products. The multiplex PCR targeted two genes, and the protocol utilized enrichment steps and required laborious gel-based analysis of PCR-amplified products. On the other hand, the 5′ nuclease PCR assay (TaqMan assay) is a relatively new PCR nucleotide sequence detection system that uses the 5′ nuclease activity of Taq DNA polymerase and detects the PCR-amplified products by hybridization and cleavage of a double-labelled fluorogenic probe during the amplification. In hogs *Y. enterocolitica* was not detected by culture methods, but was found by both multiplex PCR (7%) and the TaqMan assay (55%). In processed pork products such as chitterlings, the pathogen was detected by the culture method (8%) and the multiplex PCR (27%), but the TaqMan assay was more sensitive (79%). In ground pork the pathogen was not isolated by the culture method but was detected by the multiplex PCR (10%) and the TaqMan assay (44%). It is of interest to note that the pathogen was more frequently detected in ground pork and chitterlings than in freshly slaughtered hogs, which suggests post slaughter contamination might have occurred. These results demonstrated that the TaqMan assay was several-fold more sensitive than the multiplex PCR assay and the bacteriological culture method. The findings further emphasized that methodological factors can markedly affect results. The TaqMan assay has also been applied to the detection of *Salmonella* (Chen et al. 1997), *E. coli* O157:H7 (Oberst et al. 1998), *Listeria* (Bassler et al. 1995), and *Cryptosporidium parvum* (Kruger et al. 1998) but with varying sensitivities. For example, the 5′ nuclease detection system could detect *E. coli* O157:H7 when $\geq 10^3$ and $\geq 10^4$ CFU/ml were present in pure culture and ground beef-broth mixture, respectively, whereas the detection limits for *Salmonella* and *L. monocytogenes* were around 2 and 50 CFU in pure cultures, respectively. As for the detection of *C. parvum* in environmental water samples, TaqMan and PCR detection methods shared a similar sensitivity with the conventional microscopy method but results were inconsistent. It was believed that the contaminating debris in the water interfered with DNA extraction.

Traditionally, differentiation of *C. jejuni* from other *Campylobacter* spp. including *C. coli* relies on its hydrolysis of hippurate but it was shown that this is not always a dependable reaction (Rautelin et al. 1999). In this case, PCR is a more reliable method than culture techniques and ELISA because it can be *C. jejuni* or *C. coli* specific. The latter two techniques detect all thermotolerant *Campylobacter* spp. but still require conventional diagnostic testing for species identification which is time consuming. A study by Lilja & Hanninen (2001) showed that a commercial automated ELISA method took 2.5 days for detection of all thermotolerant campylobacters whereas the PCR technique required 3 days to identify *C. jejuni*. Although the PCR method was not as simple as the automated ELISA, it was species-specific. The specificity and sensitivity of the PCR technique make it ideal for the detection and identification of *Campylobacter* that is present in low numbers among competing microflora. These advantages of the amplification technique were proven in a study designated to detect small numbers of *C. jejuni* and *C. coli* cells in environmental water, sewage, and food samples (Waage et al. 1999). The semi-nested PCR assay used was able to detect 3–15 CFU of *C. jejuni* per 100 ml of water containing high background flora (10^4 CFU/ml) and ≤ 3 CFU per g of food. However, because PCR assays are based on detection of intact DNA rather than intact viable cells, the possibility of false-positive results exists due to amplification of DNA from dead or non-viable cells. An alternative amplification system termed NASBA® (isothermal transcription based nucleic acid amplification) that targets RNA sequences successfully, offers a much more rapid (32 h analysis time) identification of *C. jejuni*, *C. coli*, and *C. lari* (Uyttendaele et al. 1996). This detection method has good sensitivity; only 6 CFU of *C. jejuni* in the presence of 10^6 CFU background bacteria produced a positive signal (Uyttendaele et al. 1994). Nonetheless, this nucleic acid amplification system (that targets 16S rRNA) is also incapable of differentiating between viable and non-viable cells of *Campylobacter*.

An innovative combined PCR-ELISA method was described for the detection of *Y. enterocolitica* (Uyeda et al. 1996). This method combines the ELISA assay and PCR technique to achieve more rapid and sensitive detection compared to conventional gel electrophoresis of PCR products. The combined method was designed to detect only virulent (*ail*-bearing) strains of *Y. enterocolitica*. It is based on the immobilization of the biotin-labelled PCR product (probe and target) to the streptavidin-coated wells of a microtiter plate and the subsequent detection of the product by a fluorescein-labeled *ail* internal probe (colorimetric assay). This PCR-ELISA method allows for the processing of a large number of samples, is relatively inexpensive, eliminates the need for electrophoretic and photographic equipment, and avoids use of potential carcinogens such as ethidium bromide (used in the gel electrophoretic detection of PCR products).

The PCR technique is both sensitive and specific, but various technical difficulties can result in both false-negative and false-positive results. It may give false-negative results due to the presence of inhibitors in samples which are often co-extracted along with bacterial DNA (Holland et al. 2000). It may also give false-positive results due to non-specific amplification of DNA. The sensitivity of PCR assays may be reduced when there are small numbers of target organism or when there are a large number of other bacteria relative to the number of target organisms. These problems were reported for detection of *E. coli* O157:H7 directly from human stools (Holland et al. 2000). The effectiveness of the PCR procedure is dependent on sample chemistry since there are varying amounts and types of PCR inhibitors in different food samples. In one study, stronger inhibition was found in raw milk samples than in chicken carcass rinses and this was probably due to the higher protein and fat content of raw milk (Chen et al. 1997). These types of problems can be resolved by using a more effective DNA purification procedure.

Molecular methods can provide a high level of sensitivity and specificity, but often the major drawback is the need for extensive sample preparation and purification before the techniques can be applied. Other disadvantages include their high capital cost and the requirement for trained users. There is a need to develop sample preparation techniques that are time efficient, cost effective, that require minimum training, and are reproducible from user to user. Further, molecular methods are usually only qualitative and not quantitative. In addition, they are able to detect both viable and non-viable cells which may be a positive or negative feature. Among other limitations, molecular methods may not identify new strains of pathogens. That is, an organism can only be identified if its taxonomy is already known. Organisms (e.g. "deletion mutants") not previously isolated will not be identified if they do not carry the target gene base sequence.

Recently, a biosensor was developed at Cornell University for more rapid and sensitive detection of pathogens such as *E. coli* O157:H7, *C. parvum*, and *Listeria* (Friedlander Jr. 2002). While culture methods take days and PCR methods take several hours, the biosensor works in just minutes. The basis for this biosensor is the liposome, a microscopic cell-like structure made in the laboratory by adding an aqueous, marker-containing solution to a phospholipid mixture. On the outside membrane of this structure, specific antibodies were affixed. When the target pathogen was present, it bound to the antibodies and the liposome membrane was ruptured, releasing the dye (marker). For practical application, a paper test strip impregnated with liposomes is placed in a solution containing pathogens. The change in colour of the test strip indicates the presence of the pathogen. However, the detection limit of this biosensor was not stated or compared with other existing biosensors. Generally, successful biosensor detection of target organisms requires an enrichment step. While the cost is presently a drawback for these methods, the potential for their use as a field or production line screening test is high.

DETECTION OF VIRUSES

The standard method for the detection of enteroviruses involves the use of cell culture assay, which is expensive and time consuming (at least 3 days).

Cell culture of environmental samples is also complicated by the presence of organic and inorganic matter that is toxic to the cell (Reynolds et al. 1996). The PCR technique is a faster and more sensitive way. Decreased time, cost and increased sensitivity of the PCR procedure allow for the detection of low numbers of viruses usually found in environmental samples. The main drawback of this approach is that it is unable to distinguish between amplification of infectious and non-infectious viral sequences. Wolfaardt et al. (1995) reported a PCR detection technique for small round structured viruses including Norwalk-like viruses (NLV) in clinical and environmental samples. These viruses have been identified as a major cause of human gastroenteritis (Potter et al. 2002). Cloning and sequencing of the Norwalk virus genome were used to detect them and related viruses by a reverse transcription-polymerase chain reaction (RT-PCR). The authors found the RT-PCR procedure to be more sensitive than Norwalk virus antigen detection by an ELISA method. The procedure was proven suitable for environmental samples such as sewage, sewage sludge, river water, and tap water. Unfortunately, environmental samples can also contain substances that inhibit PCR amplification of target DNA and RNA. Reynolds et al. (1996) combined a cell culture technique with a RT-PCR procedure for the detection of infectious enteroviruses. The integrated approach used increased the volume of the environmental material sampled (equivalent to those used with the culture method) and this reduced the effects of compounds that interfered with PCR amplification.

DETECTION OF PROTOZOANS

Giardia and *Cryptosporidium* are two major protozoan pathogens of concern for a number of reasons. Their cysts and oocysts are resistant to conventional water chlorination, and can persist and remain infective for extended periods in water and the environment. They are produced in large numbers in fecal matter, are difficult to detect in water, can cross-infect different animal species, and have low infectious doses in humans (Veal 2000). Furthermore, traditional microbial indicators of water quality (i.e. total and fecal coliforms) do not correlate well with the presence of protozoans. Isolation of protozoans from environmental samples usually employs the use of a three step procedure, i.e. concentration, purification, and detection. Because these protozoans cannot be routinely cultured, current detection methods rely on direct examination of water concentrates. To achieve the sensitivity required for detection, large volumes of water (at least 1 to 100 L) are concentrated by filtration. Among commonly used filtration methods are cartridge, membrane, vortex flow or capsule filtration coupled with chemical flocculation (Faulkner et al. 2000; Kaucner & Stinear 1998). After filtration, cysts and oocysts are separated (purified) from contaminating debris using techniques such as density gradient centrifugation, density flotation, fluorescence-activated cell sorting, immunomagnetic separation, or flow cytometry (Kruger et al. 1998;). After purification, cysts and oocysts are usually detected with an immunofluorescence assay (IFA) or the PCR

Table 4.1. Advantages and disadvantages of detection methods for *Giardia* and *Cryptosporidium*.[a]

Method	Advantages	Disadvantages
Immunofluorescence assay (IFA)	• Well documented and used universally for staining protozoans • Allows numbers of cysts/oocysts present to be counted (quantitative)	• Often requires sample cleanup (eg. flotation) prior to microscopy or flow cytometry • Non-specific staining of algae and other debris • Not species-specific (such as for identification of *C. parvum* and *Giardia intestinalis*)
DAPI/PI[b] combined with IFA	• Used by a number of investigators • Detects viable *Cryptosporidium* • Well documented and can be incorporated into current US and UK methods	• Time consuming (more microscopy) • Unrelated to infectivity
Reverse transcriptase-PCR	• Can be employed directly after the concentration step (ie. no flotation) irrespective of sample turbidity • Detects viable *C. parvum* and *Giardia* spp. • Represents latest technology and method has been peer reviewed in international scientific journals • Can specifically detect *C. parvum*	• Presence/absence test only (qualitative) • Not widely used and therefore not yet subjected to trial by numerous investigators • Unrelated to infectivity

[a]Modified from O'Toole & Kaucner (1998).
[b]4',6-diamidino-2-phenylindole (DAPI) and propidium iodide (PI), two fluorogenic vital dyes.

technique. At all three stages, there is a possibility for cysts and oocysts to be lost (Veal 2000). The advantages and disadvantages of currently used detection methods for protozoans are listed in Table 4.1. Detection of protozoans in the environment is affected by factors such as the small size of cysts and oocysts, the relatively low number of protozoans in water, and the difficulty in identifying cysts and oocysts among other particles and debris (O'Toole & Kaucner 1998). Current detection methods are variable in their performance and the limitations include poor reliability, high cost, complexity of procedures (which require highly skilled users), the slow and tedious nature of the procedure, the inability to discriminate between living and dead cysts and oocysts, the inability to determine infectivity of cysts and oocysts, and the inability to differentiate between species (Table 4.1). For example, the IFA microscopy technique recognizes all species of *Cryptosporidium* because of the broad specificity of the primary antibody, but only *C. parvum* is recognized as a human pathogen. Therefore, species-specific detection is necessary to avoid overestimates due to false positive results (Rochelle et al. 1997).

Filtration or concentration methods have been reported to have poor recovery and are less effective when there is high sample turbidity (Kaucner & Stinear 1998). Chemical flocculation is effective but can inactivate viable organisms. Filtration methods must maintain cyst and oocyst infectivity which is essential for assay success (Carreno et al. 2001). The IFA method is very laborious, time consuming, and accuracy depends on user experience, which often results in reproducibility problems (Kruger et al. 1998). PCR is an attractive diagnostic procedure for detection of protozoans because it is rapid, sensitive, and species specific. However, significant sample concentration is required to reduce large sample volumes to microliter sizes that are compatible with PCR. Unfortunately, concentration of pathogens or nucleic acids in water samples also leads to concentration of inhibitory substances such as humic acids that can reduce PCR efficiency and reliability. One approach that was used to overcome the problem was the use of paramagnetic beads to purify target nucleic acids (Kaucner & Stinear 1998). The viability of cysts and oocysts can be determined by vital dye staining (such as 4′,6-diamidino-2-phenylindole and propidium iodide, commonly termed DAPI and PI, respectively), exposure to excystation solutions, or growth on cell culture (Campbell et al. 1992; Carreno et al. 2001; Quintero-Betancourt et al. 2002). New methods based on short-lived nucleic acids (e.g. RNA) such as fluorescence *in situ* hybridization (FISH) can determine viability and differentiate species of protozoans (Veal 2000). However, these techniques still do not determine the infectivity of cysts or oocysts and their likelihood of causing disease in humans. Non-infective cysts and oocysts are of little concern for disease transmission. The infectivity of cysts and oocysts can be tested by infection of mice or cell lines in tissue culture (Carreno et al. 2001). An interesting method was described by Rochelle et al. (1997) which combined cell culture assay and a RT-PCR technique to determine infectivity of *C. parvum*. Oocysts were first inoculated onto monolayers of cells and grown on microscope slides. Culture infections were then detected by *C. parvum* specific RT-PCR of extracted mRNA which targeted only the genes for heat shock proteins. A single infectious oocyst was detected by this procedure. While the animal infectivity method is tedious, difficult, expensive and not readily available for normal laboratory analysis, it is considered the only method that provides direct information about the ability of protozoans to cause disease (Neumann et al. 2000). Recently, however, studies have shown that there are 150 different species of infectious *C. parvum* and those that infect animals and humans are from different species (Abrahamsen 2002, personal commun.). These findings have put reliability and validity of animal infectivity methods in question.

Finally, different laboratories across the world use different techniques for isolation and detection, which often yields a wide variation in results. Sometimes there is complete disagreement between laboratories on whether or not protozoans are present and there is often disagreement on the measured concentrations. Further, considerable variation was noted between laboratories using the same methodology, with many laboratories reporting either false positive or negative results (Clancy et al. 1994 cited by McClellan 1998).

CONCLUSIONS

Regardless of methodology, detection and identification of pathogens from animals and the environment are complicated by many factors. They include intermittent shedding of pathogens by animals, variable sensitivities of different assays, lack of reproducibility between users and on repeated testing of animals, ambiguity of negative results, and choice of the right sample. Culture methods are standard for the detection of bacteria despite their insensitivity (inability to detect low numbers and differentiate between species), expense, and labour intensity. Their sensitivity is also affected by sample handling, type of enrichments, and the VBNC condition of some bacteria. The standard method for detection of viruses is cell culture assay and for protozoans is fluorescence microscopy. While cell culture assay is expensive and time consuming, the microscopy method is highly unspecific and success is dependent on user experience. For protozoan pathogens there is also a major problem with determination of cyst and oocyst infectivity which requires more research. Serological methods are a more sensitive approach and can identify more than one serotype of a pathogenic species. The major drawback is that an antibody response to a pathogen does not indicate an active infection in the animal or that fecal shedding is occurring. The PCR technique is the most sensitive and specific detection method available but is not widely used due to the extensive sample preparation required, significant capital cost, and requirement for trained users. Many studies have focused on refining the sensitivity of PCR techniques by purifying nucleic acids from contaminating matter that may inhibit PCR amplification or by combining PCR with ELISA to achieve a more rapid, sensitive and possibly quantitative detection (Uyeda et al., 1996). PCR is certainly a promising approach for detection of pathogens. Biological sensors are also innovative and their rapid detection times are appealing. However, because they involve the use of antibody, their specificity may be questioned as some pathogens may share similar antigens.

REFERENCES

Abrahamsen, M. S. 2002. Dept. Vet. Pathol., Univ. MN. Personal communication at Ann. Meet. Can. Soc. Microbiol. June 19, Saskatoon, SK.

Bassler, H. A., S. J. A. Flood, K. J. Livak, J. Marmaro, R. Knorr, and C. Batt. 1995. Use of fluorogenic probe in a PCR-based assay for the detection of *Listeria monocytogenes*. Appl. Environ. Microbiol. 61:3724–3728.

Baum, D. H., D. L. Harris, B. Nielsen, P. J. Fedorka-Cray, and K. Steckleberg. 1996. Comparison of serology and culture for detecting *Salmonella* infection of 5 to 7 month old swine [Online]. Iowa State University swine research report. URL: http://www.extension.iastate.edu/pages/ansci/swinereports/asl-1407.pdf (Accessed: June 18, 2002).

Beuchat, L. R. 1985. Efficacy of media and methods for detecting and enumerating *Campylobacter jejuni* in refrigerated chicken meat. Appl. Environ. Microbiol. 50:934–939.

Boyapalle, S., S. Kanuganti, I. V. Wesley, and P. G. Reddy. 1999. Comparison of a multiplex and 5′ nuclease PCR assays for the rapid detection of pathogenic *Yersinia enterocolitica* in swine and pork products [Online]. Iowa State University Swine Research Report. URL: http://www.extension.iastate.edu/ipic/reports/99swinereports/asl-1705.pdf (Accessed: June 18, 2002).

Buswell, C. M., Y. M. Herlihy, L. M. Lawrence, J. T. M. McGuiggan, P. D. Marsh, C. W. Keevil, and S. A. Leach. 1998. Extended survival and persistence of *Campylobacter* spp. in water and aquatic biofilms and their detection by immunofluorescent-antibody and—rRNA staining. Appl. Environ. Microbiol. 64:733–741.

Campbell, A. T., L. J. Robertson, and H. V. Smith. 1992. Viability of *Cryptosporidium* parvum oocysts: correlation of in vitro excystation with inclusion or exclusion of fluorogenic vital dyes. Appl. Environ. Microbiol. 58:3488–3493.

Carreno, R. A., N. J. Pokorny, S. C. Weir, H. Lee, and J. T. Trevors. 2001. Decrease in *Cryptosporidium parvum* oocyst infectivity in vitro by using the membrane filter dissolution method for recovering oocysts from water samples. Appl. Environ. Microbiol. 67:3309–3313.

Chen, S., A. Yee, M. Griffiths, C. Larkin, C. T. Yamashiro, R. Behari, C. Paszko-Kolva, K. Rahn, and S. A. DeGrandis. 1997. The evaluation of a fluorogenic polymerase chain reaction assay for the detection of *Salmonella* species in food commodities. Int. J. Food Microbiol. 35:239–250.

Crook, J., R. S. Engelbrecht, M. M. Benjamin, R. J. Bull, B. A. Fowler, H. E. Griffin, C. N. Haas, C. L. Moe, J. B. Rose, and R. R Trussell. 1998. The viability of augmenting drinking water supplies with reclaimed water. In: D. A. Dobbs (ed.). Issues in Potable Reuse. National Academy Press, Washington, D. C. pp. 121.

Davies, P. R., J. A. Funk, M. G. Nichols, J. M. O'Carroll, P. K. Turkson, W. A. Gebreyes, S. Ladely, and P. J. Fedorka-Cray. 1999. Effects of methodologic factors on detection of *Salmonella* in swine feces [Online]. Proceedings of the 3rd international symposium on the epidemiology and control of *Salmonella* in pork, August 5–7, 1999, Washington D. C. URL: http://www.isecsp99.org (Accessed: June 18, 2002).

Davies, P. R., P. K. Turkson, J. A. Funk, M. A. Nichols, S. R. Ladely, P. J. Fedorka-Cray. 2000. Comparison of methods for isolating *Salmonella* bacteria from feces of naturally infected pigs. J. Appl. Microbiol. 89:69–177.

Faulkner, B., R. Thurman, A. Champion, and D. Veal. 2000. Detecting protozoa in water: a comparison of methods [Online]. URL: http://www.eidn.com.au/ausgermproject2.htm (Accessed: July 2, 2002).

Friedlander Jr., B. P. 2002. CU-developed sensor to speed detection of food-borne pathogens [Online]. Cornell Chronicle. URL: http://www.news.cornell.edu/chronicle/02/3.28.02/detect_pathogens.html (Accessed: June 18, 2002).

Holland, J. L., L. Louie, A. E. Simor, and M. Louie. 2000. PCR detection of *Escherichia coli* O157:H7 directly from stools: evaluation of commercial extraction methods for purifying fecal DNA. J. Clin. Microbiol. 38:4108–4113.

Isaacson, R. E. 2000. Stealth *Salmonella*: problems for detecting *Salmonella* positive animals [Online]. A workshop on epidemiologic methods and approaches for food safety, October 18–19, 2000, Birmingham, Alabama. URL: http://vbms.unl.edu/wills/epiconf/text/4%20isaacson.pdf (Accessed: June 18, 2002).

Isaacson, R. E., R. M. Weigel, L. D. Firkins, and Bahnson, P. 1999. The effect of feed withdrawal on the shedding of *Salmonella* Typhimurium by swine [Online]. Proceedings of the 3rd international symposium on the epidemiology and control of *Salmonella* in pork, August 5–7, 1999, Washington D. C. URL: http://www.isecsp99.org (Accessed: June 18, 2002).

Kaucner, C., and T. Stinear. 1998. Sensitive and rapid detection of viable *Giardia* cysts and *Cryptosporidium* oocysts in large-volume water samples with wound fibreglass cartridge filters and reverse transcription-PCR. Appl. Environ. Microbiol. 64:1743–1749.

Kehl, K. S., P. Havens, C. E. Behnke, and D. W. K. Acheson. 1997. Evaluation of the premier EHEC assay for detection of shiga toxin-producing *Escherichia coli*. J. Clin. Microbiol. 35:2051–2054.

Kruger, P., A. Wiedenmann, and K. Botzenhart. 1998. Detection of *Cryptosporidium* oocysts in water: comparison of the conventional microscopic immunofluorescence method with PCR and TaqMan PCR [Online]. OECD workshop Interlaken '98 on molecular technologies for safe drinking water. URL: http://www.eawag.ch/publications_e/proceedings/oecd/proceedings/Krueger.pdf (Accessed: July 2, 2002).

Lilja, L., and M-L. Hanninen. 2001. Evaluation of a commercial automated ELISA and PCR-method for rapid detection and identification of *Campylobacter jejuni* and *C. coli* in poultry products. Food Microbiol. 18:205–209.

Louie, M., S. Read, A. E. Simor, J. Holland, L. Louie, K. Ziebell, J. Brunton, and J. Hii. 1998. Application of multiplex PCR for detection of non-O157 verocytotoxin-producing *Escherichia coli* in bloody stools: identification of serogroups O26 and O111. J. Clin. Microbiol. 36:3375–3377.

Lu, P., B. W. Brooks, R. H. Robertson, K. H. Nielsen, and M. M. Garcia. 1997. Characterization of monoclonal antibodies for the rapid detection of foodborne campylobacters. Int. J. Food Microbiol. 37:87–91.

McClellan, P. 1998. *Cryptosporidium* and *Giardia*—a picture of uncertainty [Online]. In: Assessment of the contamination events and future directions for the management of the catchment. URL: http://water.sesep.drexel.edu/outbreaks/Sydney_rpt3Ch2.pdf (Accessed: July 2, 2002).

Neumann, N. F., L. L. Gyurek, L. Gammie, G. R. Finch, and M. Belosevic. 2000. Comparison of animal infectivity and nucleic acid staining for assessment of *Cryptosporidium* parvum viability in water. Appl. Environ. Microbiol. 66:406–412.

Nielsen, B., D. Baggesen, F. Bager, J. Haugegaard, and P. Lind. 1995. The serological response to *Salmonella* serovars typhimurium and infantis in experimentally infected pigs-the time course followed with an indirect anti-LPS ELISA and bacteriological examinations. (Abs.) Vet. Microbiol. 47:205–218.

Novicki, T. J., J. A. Daly, S. L. Mottice, and K. C. Carroll. 2000. Comparison of sorbitol MacConkey agar and a two-step method which utilizes enzyme-linked immunosorbent assay toxin testing and a chromogenic agar to detect and isolate enterohemorrhagic *Escherichia coli*. J. Clin. Microbiol. 38:547–551.

Oberst, R. D., M. P. Hays, L. K. Bohra, R. K. Phebus, C. T. Yamashiro, C. Paszko-Kolva, S. J. A. Flood, J. M. Sargeant, and J. R. Gillespie. 1998. PCR-based DNA amplification and presumptive detection of *Escherichia coli* O157:H7 with an internal fluorogenic probe and the 5′ nuclease TaqMan assay. Appl. Environ. Microbiol. 64:3389–3396.

O'Toole, J., and C. Kaucner. 1998. Monitoring drinking water supplies for protozoan pathogens [Online]. 61st annual water industry engineers and operators' conference, September 2–3, 1998, Shepparton. URL: http://www.awwoa.org.au/conf_papers/1998/pdf/02_Christine_Kaucner.pdf (Accessed: July 2, 2002).

Quintero-Betancourt, W., E. R. Peele, and J. B. Rose. 2002. *Cryptosporidium parvum* and *Cyclospora cayetanensis*: a review of laboratory methods for detection of these waterborne parasites. (Abs.) J. Microbiol. Methods, 49:209–224.

Park, C. H., N. M. Vandel, and D. L. Hixon. 1996. Rapid immunoassay for detection of *Escherichia coli* O157 directly from stool specimens. J. Clin. Microbiol. 34:988–990.

Potter, M., and IFT expert report panelists. 2002. Emerging microbiological food safety issues: implications for control in the 21st century [Online]. IFT Expert Report. URL: http://www.ift.org (Accessed: February 25, 2002).

Rajic, A., and J. Keenliside. 2001. *Salmonella* in swine [Online]. Bacon Bits, vol. X, February 2001. URL: http://www.agric.gov.ab.ca/livestock/pork/baconbits/0102.pdf (Accessed: June 19, 2002).

Rautelin, H., J. Jusufovic, and M-L. Hanninen. 1999. Identification of hippurate-negative thermophilic campylobacters. (Abs.) Diagn. Microbiol. Infect. Dis. 35:9–12.

Reynolds, K. A., C. P. Gerba, and I. L. Pepper. 1996. Detection of infectious enteroviruses by an integrated cell culture-PCR procedure. Appl. Environ. Microbiol. 62:1424–1427.

Rochelle, P. A., D. M. Ferguson, T. J. Handojo, R. de Leon, M. H. Stewart, and R. L. Wolfe. 1997. An assay combining cell culture with reverse transcriptase PCR to detect and determine the infectivity of waterborne *Cryptosporidium parvum*. Appl. Environ. Microbiol. 63:2029–2037.

Scotter, S. L., T. J. Humphrey, and A. Henley. 1993. Methods for the detection of thermotolerant campylobacters in foods: results of an inter-laboratory study. J. Appl. Bacteriol. 74:155–163.

Spier, S. J., B. P. Smith, J. W. Tyler, J. S. Cullor, G. W. Dilling, and L. D. Pfaff. 1990. Use of an ELISA for detection of immunoglobulins G and M that recognize *Salmonella* Dublin lipopolysaccharides for prediction of carrier status in cattle. Am. J. Vet. Res. 51:1900–1904.

Stern, N. J., and J. E. Line. 1992. Comparison of three methods for recovery of *Campylobacter* spp. from broiler carcasses. J. Food Protect. 55:663–666.

Sutcliffe, E. M., D. M. Jones, and A. D. Pearson. 1991. Latex agglutination for the detection of *Campylobacter* species in water. Lett. Appl. Microbiol. 12:72–74.

Uyeda, J., K. Harmon, and I. Wesley. 1996. A PCR ELISA method for the detection of *Yersinia enterocolitica* [Online]. Iowa State University Swine Research Report. URL: http://www.extension.iastate.edu/Pages/ansci/swinereports/asl-1419.pdf (Accessed: June 18, 2002).

Uyttendaele, M., A. Bastiaansen, and J. Debevere. 1997. Evaluation of the NASBA® nucleic acid amplification system for assessment of the viability of *Campylobacter jejuni*. Int. J. Food Microbiol. 37:13–20.

Uyttendaele, M., R. Schukkink, B. van Gemen, and J. Debevere. 1994. Identification of *Campylobacter jejuni*, *Campylobacter coli* and *Campylobacter lari* by the nucleic acid amplification system NASBA®. J. Appl. Bacteriol. 77:694–701.

Uyttendaele, M., R. Schukkink, B. van Gemen, and J. Debevere. 1996. Comparison of the nucleic acid amplification system NASBA® and agar isolation for detection of pathogenic campylobacters in naturally contaminated poultry. J. Food Protect. 59: 683–687.

Van Zijderveld, F. G., A. M. van Zijderveld-van Bemmel, and J. Ankotta. 1992. Comparison of four different enzyme linked immunosorbent assays for serological diagnosis of *Salmonella* Enteritidis infections in experimentally infected chickens. J. Clin. Microbiol. 30:2560–2566.

Veal, D. 2000. Detection of *Cryptosporidium* oocysts and *Giardia* cysts in water [Online]. URL: http://www.bio.mq.edu.au/flowgrid/root/current_research/methods/methods.html (Accessed: July 2, 2002).

Waltman, W. D., and E. T. Mallinson. 1995. Isolation of *Salmonella* from poultry tissue and environmental samples: a nationwide survey. (Abs.) Avian Dis. 39:45–54.

Waage, A. S., T. Vardund, V. Lund, and G. Kapperud. 1999. Detection of small numbers of *Campylobacter jejuni* and *C. coli* cells in environmental water, sewage, and food samples by a seminested PCR assay. Appl. Environ. Microbiol. 65:1636–1643.

Wang, H., E. Boyle, and J. Farber. 2000. Rapid and specific enzyme immunoassay on hydrophobic grid membrane filter for detection and enumeration of thermophilic *Campylobacter* spp. from milk and chicken rinses. J. Food Protect. 63:489–494.

Wesley, I. V., S. C. Johnson, and W. Cray. 1997. Detection of *Yersinia enterocolitica* in pigs and pork products. Iowa State University Swine Research Report. URL: http://www.extension.iastate.edu/pages/ansci/swinereports/asl-1506.pdf (Accessed: June 18, 2002).

Wolfaardt, M., C. L. Moe, and W. O. K. Grabow. 1995. Detection of small round structured viruses in clinical and environmental samples by polymerase chain reaction. (Abs.) Water Sci. Technol. 31:375–382.

CHAPTER 5

Wastewater Treatment and Alternative Methods of Swine Manure Treatment and Handling

MUNICIPAL WASTEWATER TREATMENT

Conventional municipal wastewater treatment typically comprises preliminary, primary, and secondary treatments. In some cases, additional treatment termed tertiary or advanced treatment may be required for protection of public health and the environment (NRC 1996). These treatments yield treated effluent and a concentrated stream of solids in a liquid called sludge. Treated effluent can be discharged into surface water bodies, applied to croplands, or reused whereas sludge may undergo further treatment processes prior to utilization or disposal (Fig. 5.1).

Preliminary wastewater treatment usually involves screening and grit removal. Screening of raw wastewater removes large, coarse materials (e.g. rags) that can interfere with mechanical equipment. Grit removal separates heavy, inorganic solids that can settle in channels and interfere with treatment processes. Preliminary wastewater treatment has very little effect on wastewater quality. It only prepares wastewater for subsequent primary and secondary treatments.

Primary wastewater treatment typically includes gravity sedimentation which removes certain amounts of suspended solids and biological oxygen demand (BOD) from raw wastewater. About one-third of the BOD and one-half of suspended solids are removed at this stage. During the primary treatment process, some of the wastewater constituents such as nutrients, pathogenic organisms, trace elements, and potentially toxic organic compounds can settle or

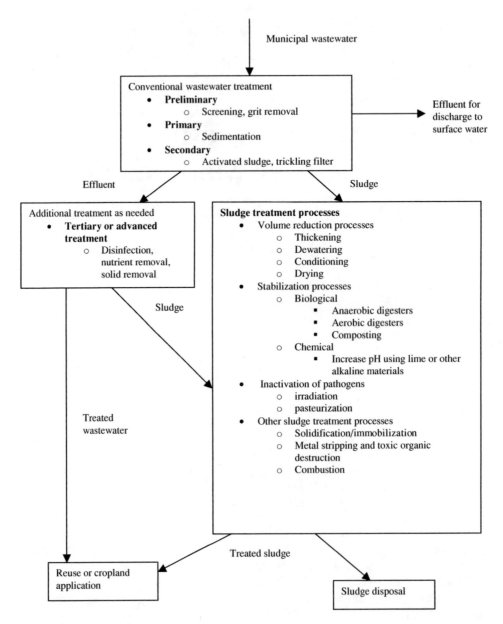

Figure 5.1. Municipal wastewater and sludge treatments. Modified from NRC (1996).

attach to suspended solids which are then removed. The residue from primary wastewater treatment is called "primary sludge".

Secondary wastewater treatment commonly employs a biological treatment process. Natural microorganisms in suspension (in an "activated sludge" process), attached to media (in a "trickling filter" process), or in ponds (such as a lagoon system) are used to degrade organic matter in the wastewater. End products of this process include carbon dioxide, trace gases, and methane which can be used to provide energy. The natural microflora forms settleable solids and following the biological treatment, the biomass is separated in sedimentation tanks and is called "secondary sludge", which is also known as "biological sludge", "activated sludge", or "trickling filter humus". Wastewater constituents that can become associated with secondary sludge include pathogens, trace elements, and organic compounds. Combined primary and secondary treatment should reduce 85% of the BOD and suspended solid concentrations in the wastewater in order to meet regulatory criteria for secondary treatment.

Conventional municipal wastewater treatment is the standard method in the U.S. for effluent discharge into surface waters. However, there are many variations in the way the treatment processes are applied. For example, primary treatment is sometimes eliminated, or long term storage in lagoons is used to replace both primary and secondary treatments.

Tertiary or advanced wastewater treatment may be used in addition to conventional municipal wastewater treatment when receiving wastewater is highly polluted or end uses require high quality effluent. The most common type of tertiary treatment is disinfection for control of pathogens (bacteria, parasites, and viruses). Disinfection of secondary effluent is commonly accomplished by chlorination. Chlorine is an inexpensive disinfectant but it can react with organic matter in wastewater and form chlorinated compounds that are undesirable for potable reuse of reclaimed wastewater. Other alternatives for disinfection include ozone and ultraviolet light treatments. These two processes do not generate undesirable products or residual disinfectant levels. Concentrations of suspended solids and BOD in secondary effluent can be further reduced by filtration or chemical coagulation. Removal of nutrients such as nitrogen and phosphorus is also a common tertiary treatment. Nitrogen can be removed by nitrification followed by de-nitrification and phosphorus is removed by microbial uptake or chemical precipitation. Sludge from tertiary treatment is usually incorporated with sludges from primary and secondary treatments.

Sludges from wastewater treatments especially primary and secondary treatments, contain settleable solids, products of microbial synthesis, non-biodegradable organic compounds, and other wastewater constituents. These sludges have a suspended solid content ranging from less than one to three percent, with the remainder mostly water. The purposes of sludge treatment processes are to reduce the water content, avoid complications (e.g. odour production) from decomposition of biodegradable matter in sludges, and reduce the levels of pathogenic organisms and viruses in sludges prior to reuse or disposal.

Reduction of sludge volume can be achieved by one of the following treatments: thickening, dewatering, conditioning, or drying (in the order of decreasing frequency of application). The detail of these processes is discussed in NRC (1996). These processes not only remove water from sludges for better efficiency of subsequent treatments, but also reduce the storage volumes and subsequent transportation costs.

Sludge stabilization is needed to minimize complications due to biodegradation of organic compounds. Sludge stabilization processes can be biological or chemical. Anaerobic digesters are most commonly used in biological stabilization of wastewater sludge and methane gas is the by-product. Aerobic digesters are sometimes used but they require oxygen or air. Composting is another aerobic biological stabilization alternative for dewatered sludge. Composting and aerobic digesters commonly operate at thermophilic temperatures (55–60°C) whereas anaerobic digestion typically takes place at mesophilic temperatures (35–37°C). Biological stabilization achieves more extensive breakdown of organic waste. Under certain circumstances, less costly alternative chemical stabilization is used. The most commonly used chemical stabilization procedure is increasing the sludge pH to 12 or higher with lime or other alkaline materials. During the process, pathogens are killed by the heat generated from the reaction between the lime and the water in sludge. This treatment of sludge also inhibits microbial degradation of organic materials which in turn prevents odour production and reduces vector (i.e. fly) attraction.

Many of the sludge volume reduction and stabilization processes can also inactivate pathogenic microorganisms and viruses in wastewater sludges. The United States Environmental Protection Agency has established two levels of pathogen reduction for sludge treatment (EPA 2002). The first is "Processes to further reduce pathogens" (PFRP) or Class A, and the second is "Processes to significantly reduce pathogens" (PSRP) or Class B. The methods that are considered PSRP are aerobic or anaerobic digestion, air drying, composting, and lime stabilization. PFRP methods that achieve a higher degree of quality than PSRP include composting at higher temperature, heat to 80°C and drying to a moisture content of 10% or lower, heat treatment to 180°C for 30 min, or thermophilic aerobic digestion. Another approach to further reduce pathogens is to apply an additional method to Class B treatment. Two methods have been developed for this purpose: beta or gamma ray irradiation and pasteurization at 70°C for 30 min or longer.

ALTERNATIVE METHODS OF SWINE
MANURE TREATMENT AND HANDLING

Swine manure can be handled in the form of a solid, slurry, or liquid, depending on the size of the operation and capital investment available. The amount of bedding materials present (such as wood shavings, sawdust, or straw) or dilution water used determines the manure form which in turn influences the manure management approach (i.e. collection, transfer, storage, and spreading techniques) (Veenhuizen et al. 1992, Fig. 5.2).

Solid manure (>15% solids) is a combination of feces, urine, and bedding with no extra water added. Semi-solid manure or slurry (4–15%) has little (if any) bedding and no extra water added. Swine manure as excreted has approximately 10% solids and can be classified as slurry (Fulhage & Pfost 2001).

Liquid manure (<4%) has sufficient water added to form a flowable mixture with no bedding. Solid manure handling is common for shed and lot systems used for swine gestation and finishing, whereas slurry and liquid manure handling are more common in modern swine production facilities (Veenhuizen et al. 1992). It is estimated that for no more than 15% of swine raised on American farms are solid manure handling systems used while around 50–60% of swine producers use slurry manure handling systems (Hatfield et al. 1998). Approximately 20–30% of the manure from swine production is processed in liquid manure systems.

Although direct land application of livestock manure is a preferred method of utilization, it is not always feasible. Normally manure is not utilized immediately, and some type of treatment is necessary. The purpose of a treatment is to convert the manure to a more stable product. Livestock manure is treated for a number of reasons: to reduce its volume and odour, to kill pathogens and weed seeds, to recover nutrients or energy, to increase its fertilizer value, or to decrease its pollution potential (Veenhuizen et al. 1992; MB 1994). One significant disadvantage of manure treatment is the loss of ammonia nitrogen from manure during the process.

In general, manure treatment processes can be categorized as physical, chemical, biological or combinations of the above. Physical treatments (may be used as primary or tertiary wastewater treatment) include simple processes such as settling, filtering, centrifuging, or drying in order to separate solids from liquid. Chemical treatments (may be used as tertiary wastewater treatment) involve addition of chemicals to condition manure. For example, addition of lime can improve dewatering characteristics of manure for better settling of solids and raise manure pH for destroying pathogens. Biological treatments (commonly used as secondary wastewater treatment) involve the use of natural microflora to change the properties of manure. Examples of biological treatment systems are anaerobic or aerobic lagoons, anaerobic digesters, and composting. While anaerobic or aerobic lagoons and anaerobic digesters are usually used for handling liquid manure, composting is more commonly used for handling solid manure. Slurry is typically stored, agitated and applied to land without undergoing any treatment (Veehuizen et al. 1992).

Anaerobic lagoons are the most common means of hog manure treatment in the U.S. (CAST 1995). Anaerobic lagoons allow conversion of swine slurry to liquid manure that is low in solids, thus allowing easier transport and application (Hatfield et al. 1998). In an ideal anaerobic lagoon system where free oxygen is absent, all organic wastes will be digested by anaerobic bacteria. Anaerobic bacteria decompose more organic matter per unit of lagoon volume than can aerobic bacteria (CAST 1995). Anaerobic digestion is often described as a two-stage process. First, one group of bacteria convert organic wastes to organic acids, then another group of bacteria convert organic acids to methane, carbon dioxide, ammonia, and hydrogen sulphide (Miner et al. 2000). Anaerobic treatment may not go to completion but can be designed to remove from 50 to more than 90% of organics in livestock waste. The methane-forming bacteria are very

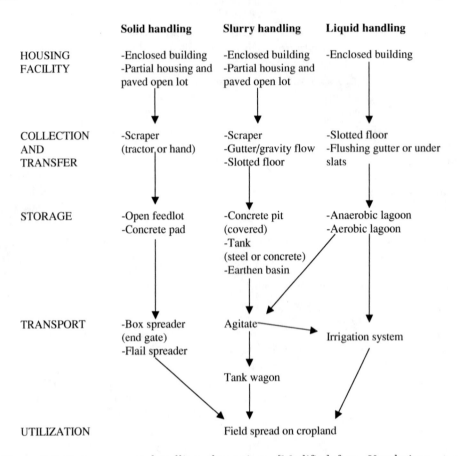

Figure 5.2. Swine manure handling alternatives. [Modified from Veenhuizen et al. (1992).]

sensitive to changes in temperature, pH, and rate of organic matter loading. Therefore, the major concern for the management of anaerobic lagoons is to maintain the right environment for methane-forming bacteria (Veenhuizen et al. 1992). A well-functioning lagoon should have a neutral pH (7 to 8) and requires a continuous loading of manure so that methane-forming bacteria receive a steady supply of waste to digest. Although anaerobic lagoons work best in warmer climates, anaerobic digestion can take place between 0 and 35°C, which is sometimes referred to as psychrotrophic digestion (Miner et al. 2000). In colder climates where below freezing temperatures are reached, anaerobic decomposition essentially ceases and the lagoon simply serves as a storage rather than treatment facility. At optimum temperature (about 35°C) it requires several months for the lagoons to digest manure, with more time being required as the temperature decreases. Because oxygen is not needed, anaerobic lagoons can be deep and have small surface area (CAST 1995). The major advantages of anaerobic lagoons are that they provide both a storage and treatment function within the same facility and that this facility operates at a minimum cost compared to other alternative treatments (Miner et al. 2000). In addition, anaerobic lagoons can also serve as a water recycling system which minimizes the storage requirement (Hatfield et al. 1998). Their limitations include offensive odour production due to incomplete decomposition of organics, loss of ammonia nitrogen in treated manure, and requirement for land area to site the lagoon. In addition to odours, potential leakage, overflow, and over-application of lagoon effluent are the major environmental concerns for anaerobic lagoons. Many large swine operations employ an anaerobic lagoon system because it provides the least-cost alternative for storage on a per-animal basis (Fulhage & Pfost 2001). This system of liquid manure handling requires less time and labour, and is flexible enough that handling can be adjusted to fit schedules for cropland application. Anaerobic lagoons are usually sized for a 365-day storage capacity.

Anaerobic digesters share the same principal two-step bacteriological processes with anaerobic lagoons. In an anaerobic digester system, the digester is covered, heated, and agitated to shorten the time required to digest manure, control odour, and collect the methane gas produced. Because anaerobic digestion is conducted in a closed tank, it conserves more nutrients by reducing ammonia volatilization and preventing odour release, both of which characterize the open anaerobic lagoons (Miner et al. 2000). Anaerobic digesters can be almost 100 times smaller than anaerobic lagoons (Veenhuizen et al. 1992). The digesters can be operated at either mesophilic or thermophilic temperatures (Miner et al. 2000). A well-controlled digester can digest manure in only 20 to 30 days. Thermophilic digestion is faster and can therefore be conducted in a smaller tank. The biogas produced by an anaerobic digester is comprised of about 50–60% methane (natural gas), 40–50% carbon dioxide, and <1% other gases such as hydrogen sulphide. Thermophilic digestion has the advantage of slightly greater biogas production than that from mesophilic digestion. Overall, anaerobic digester systems are small in size, generate limited or no odour from either the digester or treated manure, and produce biogas that can be used as an energy source. Their disadvantages include complex and expensive construction, complicated operation and requirement for daily attention, high maintenance cost (especially for thermophilic digestion), and production of explosive digester gas. Anaerobic digesters are more widely used in other parts of the world than in North America (see next section). The small-scale anaerobic digesters are commonly adopted in Asia and Western Europe where land available is limited for the construction of anaerobic lagoons.

The aerobic lagoon is an alternative to the storage and treatment of swine liquid manure. The main advantage of aerobic lagoons are that aerobic digestion tends to be more complete than anaerobic digestion and its products are more odour free (CAST 1995). In naturally aerobic lagoons, oxygen enters the lagoons by diffusion across the water surface, thus lagoons are usually shallow. Because of the need for oxygen transfer, large surface and land areas (as much as 25 times that required for anaerobic lagoons) are required for naturally aerobic lagoons. Therefore naturally aerobic lagoons are impractical and generally not recommended for hog waste treatment. Alternately, a mechanical aerator can be used in aerobic lagoons. This approach combines the odour control advantage of aerobic digestion with a relatively smaller surface area requirement. A main disadvantage of mechanically aerated lagoons is the added expense of operating the aerators. In addition, mechanically aerated lagoons have significantly reduced levels of nitrogen due to the loss of ammonia to the atmosphere. Aerobic lagoon systems require about 1 to 6 months to stabilize manure. Capital, energy, and maintenance requirements of these systems have been high enough to prevent their use from becoming popular at swine farms. Aerobic treatment is more commonly used in the municipal wastewater treatment in the form of trickling filters and activated sludges (Miner et al. 2000).

Solid manure handling is commonly employed in older swine production facilities with concrete floors (Fulhage & Pfost 2001). Solid manure is typically scraped and hauled to a field or stockpiled for later distribution when land for its application is available. Alternately, composting can be used for handling solid manure.

Composting is the aerobic biological decomposition of solid organic waste and it requires oxygen, moisture, and the right proportion of carbon to nitrogen at temperatures of 40 to 65°C (Buckley 2001; Veenhuizen et al. 1992). The rate of composting can be controlled by adjusting any of these factors. The ideal ratio of carbon and nitrogen is about 30 parts of carbon to 1 part of nitrogen. Animal products such as manure are high in nitrogen whereas plant products such as leaves, wood, and paper are high in carbon. Thus the composting process can be facilitated by mixing animal and plant waste products together. Moisture adjustment to 60% optimizes the composting process. Moisture is lost from manure during the composting process, thus it may be necessary to add water should drying occur. The optimum solids content for composting is between 40–50%, therefore it is necessary to increase the solids content of swine slurry from its normal 10% to at least 35% before it can be composted (MB 1994). During the composting process, the volume of organic waste is reduced by up to 50% but this may take one month to a year. Composted manure produces low odour, and dry compost can be easily handled. Composting still shares some of the common disadvantages of other treatments such as loss of ammonia, requirement of land area for compost piles, odour production during processing, and there is a management requirement to monitor the process (Veenhuizen et al. 1992).

Depending on the manure form, several collection, transport, and storage methods can be used. Common storage methods include under floor pits, outdoor above or below ground structures, earthen pits, lagoons, and holding ponds. Flushing gutters and scraper systems are among the methods used to collect and transport manure to storage facilities. The advantages and disadvantages of various manure management methods are listed in Table 5.1. Understandably, the most popular method for collecting, storing, hauling, and applying manure to land (which generally represents net production costs), is the one which is regionally the most cost-effective system.

SWINE MANURE HANDLING IN MAJOR
HOG PRODUCING COUNTRIES

Due to differences in climate and availability of land, capital, labour, as well as differences in management skills among producers, a single best manure handling system cannot be identified (Dickey et al. 1981). Major hog producing countries have developed different ways to deal with swine manure (Table 5.2).

Denmark adopted a centralized biogas (anaerobic digester) plant concept, in which organic wastes from several farms, industrial plants, and households are processed in one single biogas production plant. A diagram of the centralized biogas plant concept is shown in Fig. 5.3. Thus far, 20 such biogas plants have been established in Denmark and have operated successfully for over 15 years (Hjort-Gregersen 1999). The initial purpose in development of centralized waste processing plants was for energy production. It later became evident that the process made a considerable contribution to improved environmental protection by contributing to efforts with respect to agricultural, household and industrial waste recycling, and by contributing to greenhouse gas reduction. The major biomass resource for biogas production in Denmark is animal manure. Approximately 75% of the biomass treated in biogas plants is manure whereas the other 25% is waste from the food processing industry. Only a few biogas plants treat household waste. In the biogas plant, animal manure and organic waste are mixed and placed in anaerobic digester tanks for 12–25 days. Biogas produced from the digestion process is cleaned and utilized in combined heat and power production plants, where heat is later distributed to district heating systems and electricity is directed to the power grid. The digested manure is stored in slurry storage tanks in biogas plants or at nearby farms where it can be used as a fertilizer by farmers. Due to the small land base and intensive livestock production in Denmark, the laws restrict nitrogen utilization from animal manure and thus limit manure application on land. From the farmers' point of view, centralized biogas plants make it easier to meet the legislative requirements. Although biogas plants are expensive, their location within highly dense areas of livestock production makes it possible to share the capital and operating costs of the plants among several farms (Church et al. 1999). The concentrated, centralized character of the concept also means there are only short distances from farms to the processing plants, thus reducing transportation costs for farmers. High initial costs have limited the number of biogas digesters in North America (CAST 1995). In countries where fuel is very expensive and not readily available, the energy recovered from biogas production is worth the investment.

The Netherlands has some of the strictest manure handling regulations among pork-producing countries (PP 1997). In this country, manure management practices are strictly regulated and laws are enforced to reduce odour and pollution problems. Environmental concerns in the Netherlands have severely affected its livestock production. In fact, the Dutch government has implemented programs to reduce its pig and other livestock production in order to reduce the nutrient surplus from animal manure production (Church et al.

Table 5.1. Advantages and disadvantages of alternative manure management systems.[a]

Method	Engineering considerations	Advantages	Disadvantages
Under floor storage (slurry handling)	• Design dependent on soil depth and drainage • Volume based on storage time required • Pit access for equipment • Agitation requirements • Pit ventilation	• Easy collection and storage • Minimum volume • Maximum fertilizer value	• Odours and gases • Solid accumulation • Solid agitation and removal problems
Outdoor storage (slurry handling)	• Conveyance from building to storage • Cold weather operation • Agitation requirements • Above or below ground earthen structure	• Manure gases in building minimized • Adaptable to liquid/solid separation and methane production • Maximum fertilizer value	• Extra cost for storage and transfer • Dependence on transfer system • Solid removal
Mechanical scraper (slurry handling)	• Length and width of scraper surface • Power requirement • Cable or chain unit • Layout for efficient use of equipment • Cold weather operation	• Positive removal • Handle in slurry form	• Higher cost • Equipment and time dependency • Cold weather, ice • Possible disease and drug transmission in open gutter • Maintenance • Ammonia in building
Flushing—open gutter (liquid manure handling)	• Slope, length, width, and cross section of gutter • Flush volume and frequency • Plumbing and pump selection • Flush mechanism • Recycle or fresh water • Lagoon requirements	• Lower construction cost • Quick manure removal • Less odour within building • Manure movement aided by animal access • Animals attracted to gutter, good dunging patterns	• Cleanliness dependent on proper design • Possible disease and drug transmission • Lagoon requirements • Equipment dependency
Flushing—below slat (liquid manure handling)	• All of those with open gutter • Equipment for greater flushing action required	• Retrofit to existing buildings • Less odour and ventilation requirements • Minimized possible disease and drug transmission	• Higher cost than open gutter • Cleanliness dependent on design • Lagoon requirements • Equipment and time dependency

Table 5.1. (*Continued*)

Method	Engineering considerations	Advantages	Disadvantages
Anaerobic lagoons (liquid manure handling)	• Volume, depth and shape requirements • Distance to neighbouring residences • Distribution to irrigable land • Organic loading rate • Dilution water availability	• Storage and application flexibility • Low-solid liquid for simple irrigation and recycle for flushing system • Low cost and labour	• Land requirement • Odour potential • Nitrogen loss • Sludge build-up • Recycle salt problems
Open lot (solid manure handling)	• Runoff control system • Manure storage area • Dewatering equipment for holding pond • Pen slope and accessibility for scraping	• Low cost and management • Nutrient retention in solids • Easily constructed	• High labour • Liquid and solid handling equipment • Cold weather effects on pigs and producers

[a]Modified from Dickey et al. (1981).

1999). Due to the stringent nitrogen and phosphorous regulations, farmers load manure on tanker trucks and ship it for long distances to other 'manure-poor' countries such as northern France, Italy, or Africa. While this system enhances the nutrient content of poor quality soil and improves potential crop production in remote locations, it results in high transportation costs.

Intensive hog raising in Taiwan produces a remarkable 730 million tons of swine manure each year (FFTC 2001). The extremely small land area of Taiwan makes land application of manure for nutrient recovery totally unsustainable. Due to the environmental concerns and high density of its population, the country adopted expensive waste treatment systems similar to the municipal sewage treatment systems used in North America (Church et al. 1999). The full-blown engineered system which includes aerobic treatment is expensive in terms of capital and operating costs. The system used in Taiwanese hog barns also combines liquid flushing with manure treatment. The flush water is purified and discharged into a river or recycled as flushing water in the barns. The solid waste is transported to central composting sites to be converted into organic fertilizers (Beghin & Metcalfe 1998). The government has been actively involved in the improvement of waste treatment processes. Its current goal is to achieve a discharge effluent with a BOD of 80 mg per liter and a total suspended solids concentration of 200 mg per liter (Miner et al., 2000). In the future, it expects to achieve a level of 20 and 100 mg per liter for BOD and total suspended solids, respectively. These ambitious levels of pollutant removal will continue to make the wastewater treatment more expensive. Nonetheless in this system, swine

Table 5.2. Swine manure handling systems in major hog producing countries.[a]

Country	Swine manure handling system
Denmark	Centralized manure biogas generating facilities • anaerobic digester system • produce fertilizer, heat and power • *expensive* system due to capital and operating costs
Taiwan	Full-blown engineered waste handling and treatment systems • similar to municipal sewage treatment used in N.A. • combine liquid flushing systems in the hog barns and manure treatment systems that purify water • allow water reuse or return to surface water • *expensive* system due to capital and operating costs
Holland	Export of manure to "manure-poor" regions • such as northern France, Italy, or Africa • helps to enhance traditionally poor soil quality and nutrient content to improve crop production • *expensive* system due to long distance transportation
United States	Lagoon storage and treatment • land application via irrigation systems for nutrient recovery • lack of local land base to spread the manure or properly utilize the nutrients • manure spreading lands are often over-saturated from surplus liquids and nutrients • *economical* but suffers from odour production
Canada	Earthen manure storage • similar to anaerobic lagoon but without treatment • land application • *low capital cost* but suffers from odour production

[a]Modified from Church et al. (1999).

manure is not directly spread on cropland and thus probably does not affect public health.

In North America, swine manure is commonly handled as a slurry or liquid, stored in large earthen basins or lagoons, and this is followed by direct land application via irrigation for nutrient recovery and crop yield improvement. The U.S. hog industry annually generates approximately 100 million tons of fresh manure (CAST 1995). Anaerobic lagoons are commonly used for swine manure treatment in the U.S. Lagoons are usually single cell and designed based on the organic matter loading and the regional climate (Liu et al. 2001). Lagoon liquids are recycled for flushing, applied to land, or discharged into a water body. Hog production in the United States is usually concentrated within specific small areas due to the availability of feed, water and suitable climate (Church et al. 1999). This concentrated nature means that it also suffers from the same problems experienced by the smaller, densely populated countries such as the lack of land area for manure application. Manure is usually spread to lands nearby hog production operations and this has created problems of over-saturation of soil nutrients. Although the lagoon system is economical, it produces strong

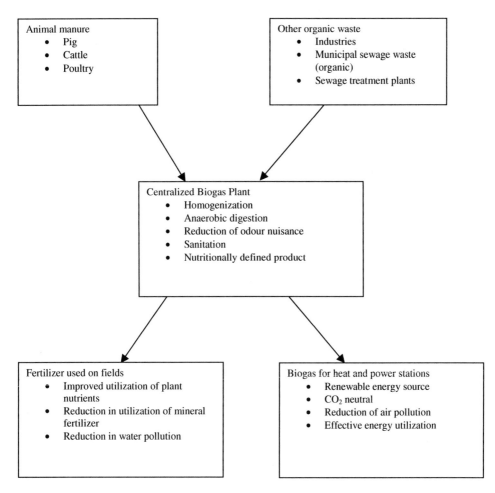

Figure 5.3. The centralized biogas plant concept in Denmark. Modified from Al Seadi (2000).

odours due to the large evaporation surfaces. In addition, it has been blamed for causing water pollution when overflow or seepage occurs (Fleming et al. 1999).

Canada shares a similar manure handling system with the U.S. While it is called lagoon storage in the U.S., it is referred to as earthen manure storage (EMS) in Canada where the extent of its digestion may be variable and is not standardized (Liu et al. 2001). A lagoon is carefully designed and managed to maintain optimum loading rates, retention time and temperature to optimize organic decomposition, whereas an EMS is simply a basin designed to store the manure between periods of land application (MB 1994). There is no specific intent to use EMS as a treatment facility. As it only serves to store manure, the only consideration for its design is the holding volume. Minimum requirement for its holding volume is about 6 months, but most EMS are designed to provide one year of storage capacity so that they can be emptied semi-annually or annually for land application. EMS employed by hog farms are usually anaerobic by default, although surface aeration does occur and provides oxygen to the top layers of manure in an uncovered EMS (Liu et al. 2001).

Uncovered EMS is common in Manitoba (MB 2001). Anaerobic digestion works best in warm climates, but the climate in Manitoba is extreme with temperature ranging as high as 40°C in the summer and as low as −40°C in the winter. Summer in Manitoba is not a problem for anaerobic digestion but winter can retard the activity of anaerobic bacteria (Liu et al. 2001). Therefore, for more efficient anaerobic treatment and potential biogas collection, covers have been suggested for EMS in Manitoba. While the advantage of an EMS is its low capital cost, its disadvantages include the large surface area for odour emission, the nutrient loss, and the maintenance requirement (MB 2001). Covered EMS can reduce odour emission and nutrient loss. Other than EMS, less common manure storage structures that are used in Manitoba include covered concrete or steel structures, solid manure storage structures, or alternative housing systems that utilize deep straw bedding for liquid absorption. Compared to EMS, concrete and steel structures are more costly whereas alternative housing systems involve lower capital cost but higher labour (MB 2001).

A study (Danesh et al. 2000) was performed to evaluate the efficiency of the Taiwanese manure treatment technology under Manitoba climate and farming conditions. The technology comprises a sequence of solid-liquid separation, anaerobic digestion, and aerobic treatment. The results showed that the technology did not produce an effluent that was suitable for discharge to surface water, as is the case in Taiwan. Differences in operational temperature were suggested to be the main reason for the loss in efficiencies. So far, other manure treatments that have had little application in Manitoba include aerobic storage, composting, anaerobic lagoons, and anaerobic digesters (MB 1994). Aerobic storage of manure is impractical due to the high cost of the mechanical aerator and its maintenance. In addition, aerobic treatment is usually needed in areas with limited land for manure application and where reduction of the nitrogen content of manure is required. However for most producers in Manitoba, there is a large land area available for all manure produced. At present, composted hog manure has a very limited market in Manitoba and composting cost may not be recovered from its sale. Anaerobic lagoons are more successful in warmer climates, but in Manitoba low temperatures are dominant for much of the year. Anaerobic digesters require high capital costs and constant management which are costly compared to the value of methane gas produced. Thus it is of little value in Manitoba where electricity is less expensive. Presently in the province since feed and energy costs are relatively low, the most cost-effective treatment system remains storage of the manure, followed by its spreading on cropland. In the future, circumstances may arise (over-application of manure to land) such that another method of treatment may become desirable.

REFERENCES

Al Seadi, T. 2000. Danish centralized biogas plants-plant descriptions [Online]. ManureNet, Agriculture and Agri-Food Canada. URL: http://res2.agr.ca/initiatives/manurenet/download/denmark_biogas.PDF (Accessed: July 15, 2002).

Beghin, J., and M. Metcalfe. 1998. Environmental regulation and competitiveness in the hog industry: an international perspective [Online]. URL: http://www.econ.iastate.edu/research/webpapers/NDN0011.pdf (Accessed: September 5, 2002).

Buckley, K. E. 2001. Composting solid manure [Online]. Agriculture and Agri-Food Canada. URL: http://www.gov.mb.ca/agriculture/livestock/pork/swine/pdf/bab14s04.pdf (Accessed: July 15, 2002).

Council for Agricultural Science and Technology (CAST). 1995. Swine production—waste management and utilization. In: Waste management and utilization in food production and processing. Task force report, No. 124, October. pp. 42–54.

Church, T., B. MacMillan, and D. Fitzgerald. 1999. Research dollars. Hoof Prints Newsletter, May. URL: http://www.agric.gov.ab.ca/livestock/hoofprints/9905.html (Accessed: May 17, 2002).

Danesh, S., D. Small, and D. Hodgkinson. 2000. Pilot plant demonstration of a three-stage waste treatment technology [Online]. DGH Engineering Ltd, MB. URL: http://res2.agr.ca/initiatives/manurenet/download/pilot_treat_dgh_2000.pdf (Accessed: July 15, 2002).

Dickey, E. C., M. Brumm, and D. P. Shelton. 1981. Swine manure management systems [Online]. URL: http://www.ianr.unl.edu/pubs/wastemgt/g531.htm (Accessed: July 5, 2002).

Environmental Protection Agency (EPA). 2002. Appendix B to Part 503-pathogen treatment processes [Online]. 40 CFR (Code of Federal Regulations) Chapter 1—Part 503. URL: http://www.access.gpo.gov/nara/cfr/cfrhtml_00/Title_40/40cfr503_00.html (Accessed: August 6, 2002).

Fleming, R., J. Johnson, and H. Fraser. 1999. Leaking of liquid manure storages-literature review [Online]. Ontario pork. URL: http://res2.agr.ca/initiatives/manurenet/download/fleming_leakmanure.pdf (Accessed: July 15, 2002).

Food & Fertilizer Technology Center (FFTC). 2001. Compost technology-new developments in composting [Online]. Newsletter 132, June. URL: http://www.agnet.org/library/data/nl/nl132/nl132.pdf (Accessed: July 15, 2002).

Fulhage, C., and D. Pfost. 2001. Swine manure management systems in Missouri [Online]. University of Missouri, MU extension, Agricultural publication EQ350—New May 1. URL: http://muextension.missouri.edu/xplorpdf/envqual/eq0350.pdf (Accessed: July 8, 2002).

Hatfield, J. L., M. C. Brumm, and S. W. Melvin. 1998. Swine manure management. In: R. J. Wright, W. D. Kemper, P. D. Millnen, J. F. Power, and R. F. Korcak (eds.). Agricultural uses of municipal, animal, and industrial byproducts. USDA Agricultural Research Service, Conservation Research No.44. URL: http://res2.agr.ca/initiatives/manurenet/download/ag_use_ars.pdf (Accessed: July 5, 2002).

Hjort-Gregersen, K. 1999. Centralised biogas plants-integrated energy production, waste treatment and nutrient redistribution [Online]. ManureNet, Agriculture and Agri-Food Canada. URL: http://res2.agr.ca/initiatives/manurenet/download/denmark_99.pdf (Accessed: July 15, 2002).

Liu, C., D. Small, and D. Hodgkinson. 2001. The effects of earthen manure storage covers on nutrient conservation and stabilisation of manure [Online]. DGH Engineering Ltd, MB. URL: http://res2.agr.ca/initiatives/manurenet/download/dgh_earthen_man_covers.pdf (Accessed: July 5, 2002).

Manitoba (MB). 2001. Farm practices guidelines for hog producers in Manitoba [Online]. Manitoba Agriculture and Food. URL: http://www.gov.mb.ca/agriculture/livestock/pork/swine/bah00s00.html (Accessed: July 8, 2002).

MB. 1994. Farm practices guidelines for hog producers in Manitoba. Agricultural guidelines development committee in cooperation with Manitoba Pork Est. Manitoba Agriculture. pp. 41–45.

Miner, J. R., F. J. Humenik, and M. R. Overcash. 2000. Managing livestock wastes to preserve environmental quality. Iowa State University Press, Ames. pp. 40–41, 135–170.

National Research Council (NRC). 1996. Municipal wastewater and sludge treatment. In: Use of reclaimed water and sludge in food crop production. National Academy Press, Washington, D. C. pp. 45–55.

Pork Producer (PP). 1997. Manure laws rule in the Netherlands [Online]. Fall issue, vol. 5, no. 3. URL: http://www.agpub.on.ca/pork/fall97/porkfl2.htm#man (Accessed: July 15, 2002).

Veenhuizen, M. A., D. J. Eckert, K. Elder, J. Johnson, W. F. Lyon, K. M. Mancl, and G. Schnitkey (eds.). 1992. Ohio livestock manure and wastewater management guide [Online]. Ohio State University Bulletin Extension, Bulletin 604. URL: http://ohioline.osu.edu/b604/index.html (Accessed: July 10, 2002).

CHAPTER 6

Pathogen Transport in the Environment and its Relation to Public Health

Land application of manure is a common practice among livestock farms in Canada. In fact, the practice has been traditionally accepted and encouraged as one component of a sustainable agricultural system. From the farmers' point of view, recycling manure not only efficiently reduces disposal problems and decreases production costs, but also recovers valuable nutrients and improves crop yields. Ideally, under proper management, this practice represents efficient utilization of animal waste with little harm to the environment. Apart from its contribution toward excess nutrients, livestock manure can also contain various types of human pathogens. Depending on the sensitivity of pathogens to environmental stress, application of manure to land may result in contamination of soil and water supplies with undesirable organisms and thus pose a health hazard to the general public.

Following the application of animal waste to land, the fate of human pathogens depends principally on their ability to survive and the opportunity for transport in those environments. The transport of microorganisms from soil to aqueous systems may occur via one of two routes, depending on soil type and conditions (Mawdsley et al. 1996). In permeable soils, the most likely transport route is down through the soil profile to land drains (e.g. field tile drainage) or groundwater. This movement is also described as leaching and essentially represents vertical movement of microorganisms. On the other hand, in saturated or impermeable soils, water and microorganisms applied may be lost through surface runoff or by flow along and above an impermeable substrate (horizontal movement). Many factors can affect the rate and extent of horizontal and vertical transport of microorganisms through soil (Tables 6.1 & 6.2), and they have been extensively studied and reported in the literature (Bitton et al. 1999; Mawdsley et al. 1995; Abu-Ashour et al. 1994).

Table 6.1. Factors that affect horizontal and vertical movement of microorganisms in soil.[a]

Movement	Factor
Horizontal	Rainfall and its intensity
	Irrigation
	Topography of land and proximity to pollutant source
	Agricultural practice
	Weather or season at time of application
Vertical	Rainfall and its intensity
	Irrigation
	Soil type or structure
	Tillage
	Soil moisture
	Temperature
	pH
	Mesofaunal activity
	Surface properties of microorganisms
	Presence of plant roots

[a]Modified from Mawdsley et al. (1995).

Shortly after manure application, most microorganisms are retained at or near the soil surface (Khaleel et al. 1980). Surface runoff as a result of rain, snowmelt or irrigation, can carry these microorganisms significant distances downstream to ground and surface waters (Abu-Ashour & Lee 2000). Surface runoff losses of microorganisms increase with rainfall and application rate, but decrease with time after application of manure and the clay content of soil (Reddy et al. 1981). Clay content of soil increases retention of microorganisms by soil particles. Generally, fine-textured soils retain microorganisms including viruses more effectively than sandy soils, because the clay mineral fraction displays a high sorptive capacity toward microorganisms due to its high surface area and ion-exchange capacity (Gerba & Bitton 1984). It is generally agreed that microbial movement is more significant under saturated (where water fills all the pores in soil) than unsaturated (where water flows only through small pores or is retained as a film around soil particles) flow conditions (Gerba & Bitton 1984). Thus, unsaturated flow allows microorganisms to get closer to particle surfaces and increases adsorption by soil particles. Leaching of pathogens through soil is primarily affected by soil structure and the velocity of water flow (e.g. rainfall or irrigation intensity) (Smith et al. 1985). Rain can rapidly move bacteria from the soil surface through soil into ground water (Howell et al. 1995). Therefore rainfall intensity is positively related to the transport distance possible for microorganisms in soil. Macropores are a characteristic feature of soil structure. They exist in soil because of old root channels, insect and animal burrows, and natural granular structure. Macropore flow allows water and microorganisms to rapidly move through an intact (non-tilled) soil structure (Smith et al. 1985). This is also described as preferential movement in well-structured soils (McMurry et al. 1998; Bundt et al. 2001).

Table 6.2. Summary of the main factors governing the transport of microorganisms through soils.[a]

Factor	Comments
Soil type	Fine-textured soils retain microorganisms more effectively than light-textured soils. Iron oxides increase the adsorptive capacity of soils. Muck soils are generally poor virus absorbents
Filtration	Straining of bacteria at soil surface limits their movement
pH	Generally, adsorption increases when pH decreases
Cations	Adsorption increases in the presence of cations (cations help reduce repulsive forces on both microorganisms and soil particles). Rainwater may desorb viruses from soil owing to its low conductivity
Soluble organics	Generally, soluble organics compete with microorganisms for adsorption sites. Humic and fulvic acids reduce virus adsorption to soils
Microorganisms	Adsorption to soils varies with type of microorganisms and strain. Generally, larger cells move faster than smaller ones
Flow rate	The higher the flow rate, the lower the microbial adsorption to soils
Saturated versus unsaturated flow	Microbial movement is less under unsaturated flow conditions

[a]Modified from Gerba & Bitton (1984).

It appears that microorganisms in the environment predominantly travel the same pathways as water. McMurry et al. (1998) applied poultry manure on intact soil blocks followed by irrigation and observed that fecal coliforms appeared where most drainage flowed. Mawdsley et al. (1996), who studied horizontal movement of *Cryptosporidium* in soil, suggested that oocysts transported in runoff stayed in the aqueous phase and did not precipitate out onto the soil surface. In addition, first rains after manure application often were shown to yield leachate with the worst bacteriological quality (Edwards & Daniel 1994; Ogden et al. 2001). Therefore rainfall is a major factor affecting both vertical (leaching) and horizontal (runoff) movement of microorganisms in soil (Abu-Ashour & Lee 2000).

Agricultural practices may also influence transport of microorganisms in soil. Because tillage disrupts structure and pores in soil, tilled soil has fewer macropores and thus retards bacterial movement better than intact, well-structured soils (Smith et al. 1985; McMurry et al. 1998). Nonetheless, one laboratory study showed that tillage practice and soil type affected but did not prevent leaching of *E. coli* O157:H7 in soil (Gagliardi & Karns 2000). Time of manure application is another factor that affects the transport behavior of pathogens in the environment. In a field study conducted in Ontario, Culley & Phillips (1982) found that manure applications in winter resulted in significantly higher fecal coliform and fecal *Streptococcus* numbers in surface runoff, and higher fecal *Streptococcus* counts in subsurface discharge when compared with applications made during other seasons. This was probably due to the longer survival of fecal bacteria at lower (winter) than higher (spring or fall) temperatures (Reddy et al. 1981). This is likely one of the reasons that some jurisdictions (including Manitoba) prohibit spreading manure on fields in winter months.

How far can pathogens travel in the environment? Studies on point source contamination using septic tank and municipal sewage treatment effluents found that coliforms moved through soil from 0.6 to 830 m depending on the soil structure (Table 6.3). Viral migration up to 408 m has been reported (Keswick & Gerba 1980 cited by Abu-Ashour et al. 1994) although most studies showed that viruses do not travel as far as bacteria (Tables 6.3 & 6.4). This is probably due to the much smaller size of viruses than bacteria, plus other factors (Gerba & Bitton 1984). For example, sludge-associated virus particles easily become trapped at the soil surface and thus their migration through the soil is limited (Bitton 1999). Public health risks associated with land disposal of animal wastes have been evaluated extensively in reviews of factors controlling the transport pattern of viruses through soil (Gerba & Bitton 1984; Mawdsley et al. 1995; Abu-Ashour et al. 1994). These factors include soil type, pH, organic matter, cations, flow rate, degree of pore saturation, and virus type (Table 6.2).

Agricultural point source pollution arises from manure contamination originating in animal feedlots, animal housing facilities, and manure storage areas, such as lagoons (Gagliardi & Karns 2000). Non-point source pollution includes manure contamination resulting from pastured animals, from roaming wild animals, or from manure spread onto fields as fertilizer or waste. A point source can lead to non-point source manure or pathogen contamination by runoff or leaching that spreads to fields and water supplies. Non-point source pathogen contamination resulting from animal manure application to fields has also gained attention recently. In the U. S., agricultural runoff influenced by non-point source pollution frequently exceeded the EPA standards for bacterial contamination of primary contact water (200 fecal coliforms per 100 ml) (Howell et al. 1995). A study of well contamination on Ontario farms found that the percentage of coliform-contaminated wells decreased significantly with increased separation of the well from the feedlot or exercise yard on livestock farms (Goss et al. 1998). In a series of field studies conducted over two years, Lee et al. (1998) reported that significant numbers of labeled nalidixic acid-resistant (NAR), *Escherichia coli* infiltrated through the soil ≥ 0.9 m deep and traveled through sub-surface tile drains to the receiving water, even though liquid manure was spread using practices accepted in Ontario. In a recent study, Ogden et al. (2001) reported leaching losses of 0.2–10% of *E. coli* after cattle slurry application on drained plots in both grassland and arable stubble.

The transport distance possible for pathogens is also influenced by the slope of the soil surface. Abu-Ashour & Lee (2000) observed that on a soil (clay loam) plot with a slope of 2%, labeled *E. coli* (NAR) traveled in soil and surface runoff for about 20 m downstream from the center of the plot. On a second plot (same soil type) with a steeper slope (6%), labeled *E. coli* was found 35 and 30 m downstream in soil and runoff, respectively. Therefore, the greater the slope the further bacteria were able to migrate in the environment.

Table 6.3. Maximum observed movement of bacteria through soil.[a]

Nature of pollution	Organism	Soil type	Travel distance (m)	Travel time (h)
Canal water on percolation beds	E. coli	Sand dunes	3.0	—
Sewage introduced through a perforated pipe	Coliforms	Fine-grained sands	1.8	—
Oxidation pond effluent	Coliforms	Sand-gravel	830	—
Secondary sewage effluent on percolation beds	Fecal coliforms	Fine loamy sand to gravel	9.1	—
Diluted settled sewage into injection well	Coliforms	Sand and pea gravel	30	35
Tertiary treated wastewater	Coliforms	Fine to medium sand	6.1	—
Tertiary treated wastewater	Fecal coliforms and streptococci	Coarse gravel	457	48
Lake water and diluted sewage	Bacillus stearother- mophilus	Crystalline bedrock	29	24–30
Primary and treated sewage effluent	Coliforms	Fine sandy loam	0.6–4.0	—
Secondary sewage	Coliforms	Sandy gravels	0.9	—
Primary sewage in infiltration basins	Fecal streptococci	Silty sand and gravel	183	—
Sewage trenches intersecting ground water	Bacillus coli (now E. coli)	Fine sand	20	27 w[b]
Sewage in pit latrine intersecting ground water	Bacillus coli	Fine and coarse sand	24	—
Sewage in bored latrine intersecting ground water	Bacillus coli	Sand and sandy clay	11	8 w
Sewage in pit latrine intersecting groundwater	Bacillus coli	Fine and medium sand	3.1	—

[a]Modified from Gerba et al. (1975) and Hagedorn (1984).
[b]w = weeks.

In spite of the constraints described, microbial contamination of sub-surface water can occur at a rapid rate. Dean & Foran (1992) used NAR E. coli and rifampicin resistant Streptococcus faecalis to trace manure movement in soil and through tile drainage. They detected labeled organisms in both soil and water, with bacterial contamination of water occurring within 20 min to 6 h following manure application. Evans & Owens (1972) observed a 30–900 fold

Table 6.4. Virus removal by soil.[a]

Nature of fluid	Virus	Soil type	Travel distance (m)	Removal by soil (%)
Spring water	Coxsackie	Garden soil	0.9	50
Spring water	T4	Garden soil	0.6	22
Sewage effluent	Polio 1	Hawaiian soil high in iron oxide	1.2	—
Lime-treated secondary effluent	Polio 1	Sand, 0.65 mm diameter	0.2	82–99.8
Tap water	Coxsackie	Sand	0.7	0–>90
Tap water	Polio 1	Unsaturated dune sand	0.6	99.5–99.9
Oxidation tank effluent	Polio 3	Sand, sandy loam, and garden soil	0.3	—
Tap water	Polio 1	Coarse and fine Ottawa sand	0.6	1–>98
Distilled water with added salts	T1, T2, and f2	Soils from Arkansas and California	0.4–0.5	>99
Distilled water	Polio 2	Low humic latersols	0.04–0.15	96–99.3
Distilled water	T4	Low humic latersols	0.04–0.15	100
Distilled water	Polio 2	Tantalus cinder	0.15–0.38	22–35
Distilled water	T4	Tantalus cinder	0.15–0.38	100
Distilled water 10^{-3} N (Ca and Mg salts)	Polio 1	Dune sand	0.2	27–44
Distilled water 10^{-5} N (Ca and Mg salts)	Polio 1	Dune sand	0.2	99.8–99.9
Distilled water	Polio 1	Sandy forest soil	0.2	97
Distilled water	T7	Sandy forest soil	0.2	88
Secondary treated sewage	Polio 1	Sandy forest soil	0.2	98.6
Secondary treated sewage	T7	Sandy forest soil	0.2	99.6

[a]Modified from Gerba et al. (1975).

increase in fecal bacterial concentration in the tile effluent from a sandy clay loam pasture field within 2 h of spreading liquid manure. It was later suggested that the rapid movement of bacteria was a result of the preferential flow through continuous macropores (Abu-Ashour et al. 1994).

In a laboratory study, Smith et al. (1985) showed that 96% of *E. coli* irrigated onto a soil column 280 mm deep were recovered in the effluent. Gannon et al. (1991) examined 19 strains of bacteria for degrees of transport in a 50 mm long soil column and obtained recovery rates ranging from 4.3–100%. Gagliardi & Karns (2000) applied *E. coli* O157:H7-contaminated manure to 100 mm disturbed soil cores followed by steady rainfall of 16.5 mm/h. The level of the pathogen in leachate they obtained was near the inoculum level for 8 h. Mawdsley et al. (1996) applied slurry seeded with *C. parvum* oocysts to soil blocks in a tilting table (800 mm L × 560 mm W × 250 mm D) followed by irrigation for up to 70 d. The oocysts were able to move both vertically and horizontally in runoff and leachate for the entire course of the experiment. McMurry et al. (1998) applied poultry manure to intact silt loam soil blocks (425 mm L × 325 mm W × 325 mm D) and obtained fecal coliforms at 10^3–10^5 CFU/ml in leachate with unsaturated flow (rainfall at 10 mm/h).

CONCLUSIONS

Thus far, both field and laboratory investigations have demonstrated that microorganisms including pathogens can migrate for significant distances and at high rates through soil in both vertical and horizontal directions. Some studies have also shown that pathogens from manure spread using currently acceptable application practices can travel through soil and reach receiving waters which are subsequently used as public water sources. It is notable that approximately 75% of the water used in Canada is taken from surface water (Goss et al. 1998). Although it is impossible to generalize and accurately predict the mobility of these pathogens in soil in Manitoba based on previous studies (due to differences in climate, topography and soil type), it is clear that the potential exists for microorganisms to be transported in the environment and contaminate regional water supplies.

Guidelines given to farmers in Manitoba have focused on the amount of manure required to provide adequate nutrients for specific crops without causing nutrient contamination of ground and surface water (AGDC 1994). Recently, revised guidelines require manure storage structures to have a minimum setback distance of 100 meters from all surface and ground water bodies (AGDC 1998). There are also recommended setback distances from water bodies for manure spreading which are dependent on the slope from the field to the water. These setback distances are calculated and designed for the prevention of nutrient contamination of water supplies, but these probably also work for prevention of pathogen contamination. However, field studies in the province should be carried out to validate this conclusion before mandated "setback" distances are used for prevention of bacterial contamination of water.

REFERENCES

Abu-Ashour, J., D. M. Joy, H. Lee, H. R. Whiteley, and S. Zelin. 1994. Transport of microorganisms through soil. Water, Air Soil Pollut. 75:141–158.

Abu-Ashour, J., and H. Lee. 2000. Transport of bacteria on sloping soil surfaces by runoff. Environ. Toxicol. 15:149–153.

Agricultural Guidelines Development Committee (AGDC). 1994. Farm practices guidelines for hog producers in Manitoba. Manitoba Agriculture. pp. 15–25.

AGDC. 1998. Farm practices guidelines for hog producers in Manitoba. Manitoba Agriculture. pp. 17–32.

Bitton, G. 1999. Public health aspects of wastewater and biosolids disposal on land. In: Wastewater Microbiology. 2nd ed. Wiley-Liss, New York. pp. 429–447.

Bundt, M. F. Widmer, M. Pesaro, J. Zeyer, and P. Blaser. 2001. Preferential flow paths: biological 'hot spots' in soils. Soil Biol. Biochem. 33:729–738.

Culley, J. L. B., and P. A. Phillips. 1982. Bacteriological quality of surface and subsurface runoff from manured sandy clay loam soil. J. Environ. Qual. 11:155–158.

Dean, D. M., and M. E. Foran. 1992. The effect of farm liquid waste application on tile drainage. J. Soil Water Conserv. 368–369.

Edwards, D. R., and T. C. Daniel. 1994. Quality of runoff from fescue grass plots treated with poultry litter and inorganic fertilizer. J. Environ. Qual. 23:579–584.

Evans, M. R., and J. D. Owens. 1972. Factors affecting the concentration of fecal bacteria in land-drainage water. J. Gen. Microbiol. 71:477–485.

Gagliardi, J. V., and J. S. Karns. 2000. Leaching of *Escherichia coli* O157:H7 in diverse soils under various agricultural management practices. Appl. Environ. Microbiol. 66:877–883.

Gannon, J. T., U. Mingelgrin, M. Alexander, and R. J. Wagenet. 1991. Bacterial transport through homogeneous soil. Soil Bio. Biochem. 23:1155–1160.

Gerba, C. P., and G. Bitton. 1984. Microbial pollutants: their survival and transport pattern in groundwater. In: G. Bitton and C. P. Gerba (eds.). Groundwater Pollution Microbiology. John Wiley & Sons, New York. pp. 65–88.

Gerba, C. P., C. Wallis, and J. L. Melnick. 1975. Fate of wastewater bacteria and viruses in soil. J. Irrig. Drain. Div. 101:157–174.

Goss, M. J., D. A. J. Barry, and D. L. Rudolph. 1998. Contamination in Ontario farmstead domestic wells and its association with agriculture: 1. results from drinking water wells. J. Contam. Hydrol. 32:267–293.

Hagedorn, C. 1984. Microbial aspects of groundwater pollution due to septic tanks. In: G. Bitton and C. P. Gerba (eds.). Groundwater Pollution Microbiology. John Wiley & Sons, New York. pp. 181–195.

Howell, J. M., M. S. Coyne, and P. Cornelius. 1995. Fecal bacteria in agricultural waters of the bluegrass region of Kentucky. J. Environ. Qual. 24:411–419.

Khaleel, R., K. R. Reddy, and M. R. Overcash. 1980. Transport of potential pollutants in runoff water from land areas receiving animal wastes: a review. Water Res. 14:421–436.

Lee, J. H., C. M. Reaume, H. R. Whiteley, and S. Zelin. 1998. Microbial contamination of subsurface tile drainage water from field applications of liquid manure. Can. Agric. Eng. 40:153–160.

Mawdsley, J. L., R. D. Bardgett, R. J. Merry, B. F. Pain, and M. K. Theodorou. 1995. Pathogens in livestock waste, their potential for movement through soil and environmental pollution. Appl. Soil Ecol. 2:1–15.

Mawdsley, J. L., A. E. Brooks, R. J. Merry, and B. F. Pain. 1996. Use of a novel soil tilting table apparatus to demonstrate the horizontal and vertical movement of the protozoan pathogen *Cryptosporidium parvum* in soil. Biol. Fertil. Soils, 23:215–220.

McMurry, S. W., M. S. Coyne, and E. Perfect. 1998. Fecal coliform transport through intact soil blocks amended with poultry manure. J. Environ. Qual. 27:86–92.

Ogden, I. D., D. R. Fenlon, A. J. A. Vinten, and D. Lewis. 2001. The fate of *Escherichia coli* O157 in soil and its potential to contaminate drinking water. Int. J. Food Microbiol. 66:111–117.

Reddy, K. R., R. Khaleel, and M. R. Overcash. 1981. Behavior and transport of microbial pathogens and indicator organisms in soils treated with organic wastes. J. Environ. Qual. 10:255–266.

Smith, M. S., G. W. Thomas, R. E. White, and D. Ritonga. 1985. Transport of *Escherichia coli* through intact and disturbed soil columns. J. Environ. Qual. 14:87–91.

CHAPTER 7

Environmental Legislation and the Economic Impact on Intensive Hog Rearing Operations from the Perspective of Enteric Pathogens of Concern to Human Health

CANADA

Livestock operations in Canada are primarily governed under provincial acts and regulations and/or local municipal by-laws. The Fisheries Act is the main federal act that addresses pollution from agricultural operations (OMAF 2000). The Fisheries Act states that fish-ways may not be damaged or obstructed and prohibits dumping of potentially harmful substances (including fertilizer, pesticide runoff, fuel, manure, and suspended solids) into waterways that are or may be frequented by fish. There is no federal act that regulates issues specific to intensive livestock operations (ILO). Provincial legislation and approaches to regulating ILO have been thoroughly described in previous reviews (OMAF 2000; Caldwell & Toombs 2000; AAFRD 1998).

Caldwell & Toombs (2000) summarized the key features of legislation used in regulating livestock facilities in each of Canada's 10 provinces (Table 7.1) and made some important observations from the data. (1) All provinces have adopted some form of legislation that has an impact on livestock production (e.g. Planning Act, Building Codes, Environmental Protection). However, only a few have adopted specific legislation in response to the intensification of livestock production (i.e. New Brunswick, Quebec, Saskatchewan, and more recently Alberta). This legislation (or its absence) defines the leadership provided at the provincial (e.g. Quebec) or municipal (e.g. Manitoba) levels. (2) All provinces except for Newfoundland have farm practices legislation that protects the use of accepted agricultural practices from lawsuits which allege nuisance. In recent years, Ontario and British Columbia have extended such legislation to include municipal by-laws that restrict normal and acceptable farm practices. (3) All provinces have adopted guidelines, strategies or policies to assist producers, provincial departments and municipalities with the siting of livestock facilities. In some areas, these guidelines serve as educational material (e.g. "Best Management Practices" in Ontario) while in others, they are a part of a regulatory framework (e.g. Lethbridge County in Alberta). (4) In most provinces, the Department or Ministry of Agriculture is the main provincial department involved in issues pertaining to ILO. However in Quebec, Manitoba and British Columbia, the Department of the Environment has a much greater role. (5) The use of a nutrient management plan (NMP) is common where livestock intensification has been prominent. In some regions (e.g. Quebec), the NMP is a regulatory tool while in others (e.g. Ontario), it serves as an educational tool. However, this will change in Ontario as a new nutrient management act (a.k.a. Bill 81), passed in the summer of 2002, comes into effect. More commonly, an NMP is a prerequisite to a certificate (e.g. New Brunswick) or a building permit (e.g. many municipalities in Ontario). Some provinces and municipalities may also require more detailed environmental studies as a requirement for a NMP (e.g. Manitoba).

In Canada, the definition of an ILO depends on the threshold number in terms of an AU or a LU for an operation in individual provinces (Table 7.1). As a result, the definition of an ILO also varies greatly among provinces. In Manitoba, an ILO is an operation that has greater than 400 AU. In Alberta, an ILO is any operation where the number of animals in confinement on the farm equals or exceeds the threshold of 300 AU and where the livestock are confined at a density of 43 AU per acre for greater than 90 consecutive days (OMAF 2000). In Saskatchewan, an ILO means the confining of livestock where the space per AU is <370 m². In Quebec, an ILO is an operation that has more than 75 AU. In Ontario, it is defined as any operation having ≥150 LU. Because it is not defined by the legislation, some municipalities in Ontario set their own threshold numbers for an ILO. Bill 81 will require all agricultural operations regardless of size to prepare NMPs.

There is no consistency across the country in defining an animal unit (AU) or a livestock unit (LU) (Tables 7.1 & 7.2). In Ontario, the LU is preferred as it is developed to reflect barn odour potential and is not related to manure nitrogen production. It is an equivalent value for each type of animal based on manure production and production cycle, and is used to equate all types of livestock in terms of relative odour intensity. After Bill 81 was passed, nutrient unit (NU) is being introduced for categorization of farms. It will be used for phase-in implementation of a nutrient management plan (NMP).

Table 7.1. Status of provincial legislation regulating intensive livestock operations in Canada.[a]

Legal & jurisdictional context	Newfoundland	Nova Scotia	Prince Edward Island	New Brunswick
Is there specific legislation?	No	No	No	New Livestock Operations Act
Are there provincial strategies/policies?	Draft commodity sector guidelines	Manure management guidelines	Yes ("Cultivating Island Solutions, 1999)	Yes (manure management guidelines)
Do programs focus on livestock only?	No	No	No	Yes
What is the key Provincial Department?	Agriculture	Agriculture	Agriculture	Agriculture
Are provincial approvals required?	No	No	No	Yes (license under New Livestock Operations Act)
Standards for siting intensive livestock operations (general applicability)				
Building permits/zoning	Provincial legislation, inconsistent municipal implementation	Provincial legislation, inconsistent municipal implementation	Provincial legislation, inconsistent municipal implementation	Provincial legislation, inconsistent municipal implementation
Separation distances	Yes (draft guidelines)	Yes (implementation by some municipalities)	Yes	Yes
Manure storage (structure/capacity)	draft guidelines	210 days	210 days	210–250 days (defined by legislation)
Environmental studies requirement	No	No	Yes	Optional
Mandatory public meetings/notification	No	No	Notification only	No
Nutrient Management Plans (NMP)	No	No	No	Yes (manure management guidelines)
3rd party review of submissions	—	—	—	Province reviews
Who can complete NMP?	—	—	—	Staff agrologist
Approach to enforcement	—	—	—	Complaint driven (Ag Review Comm.)
Register land for nutrient application	No	No	No	No
Prohibit winter manure application	No	No	No	No
Limitations on livestock densities/total size	No	No (yes)	No	Indirectly—NMP
Land base requirements for spreading	No	No	Yes	Yes
Thresholds defining intensive	No	No	Range 30–60 LU	Yes

Table 7.1. Status of provincial legislation regulating intensive livestock operations in Canada (continued). [a]

Legal & jurisdictional context	Quebec	Ontario	Manitoba
Is there specific legislation?	Environmental Quality Act	No (Nutrient Management Act 2002 was passed and draft regulations are being developed)[b]	No
Are there provincial strategies/policies?	Yes	Yes	Yes
Do programs focus on livestock only?	Livestock /crops	Livestock focus	Livestock focus
What is the key Provincial Department?	Environment	Agriculture	Environment/Agriculture
Are provincial approvals required?	Yes	No	Yes (NMP and lagoons)
Standards for siting intensive livestock operations (general applicability)			
Building permits/zoning	Provincial legislation, municipal implementation	Provincial legislation, municipal implementation	Provincial legislation, municipal implementation
Separation distances	Legislation	Provincial policy, municipal implementation	Provincial guideline (implemented by some municipalities)
Manure storage (structure/ capacity)	250 days – required by legislation	240 days (Provincial strategy) some municipal implementation	Dept of Environment, approvals required above 400 AU
Environmental studies requirement	Yes	No	Required by some municipalities
Mandatory public meetings/notification	No	No	Required by many municipalities
Nutrient Management Plans (NMP)	Yes	Provincial strategy, implemented by some municipalities	Yes
3rd party review of submissions	Yes (government)	Yes	Yes (government)
Who can complete NMP?	Agrologist	Farmers (some municipalities require consultants)	Farmers
Approach to enforcement	Fines—complaint driven	Local agriculture review committee (some municipalities try to enforce)	Random Audits
Register land for nutrient application	Yes	Some municipalities considering	Yes
Prohibit winter manure application	Yes	No	Yes for large operations
Limitations on livestock densities/total size	Indirectly through Agro-environmental fertilization plan	Indirectly through NMP	Indirectly through NMP
Land base requirements for spreading	Indirectly through Agro-environmental fertilization plan	NMP	NMP
Thresholds defining intensive	75 LU	150 LU	400 AU (200 AU for some municipalities)

Table 7.1. Status of provincial legislation regulating intensive livestock operations in Canada (continued). [a]

Legal & jurisdictional context	Saskatchewan	Alberta	British Columbia
Is there specific legislation?	Agricultural Operations Act	Agricultural Operations Practices Amendment Act 2001 was passed in November 2001 and took effect on January 1, 2002[c]	No
Are there provincial strategies/policies?	Yes	Yes	Yes
Do programs focus on livestock only?	Livestock	Livestock	Both (mushroom, greenhouse)
What is the key Provincial Department?	Agriculture	Agriculture	Environment, Agriculture
Are provincial approvals required?	Yes	Yes	No
Standards for siting intensive livestock operations (general applicability)			
Building permits/ zoning	Provincial legislation, municipal requirement	Yes	Provincial legislation, municipal requirement
Separation distances	Yes–where mandated by municipality	Yes	Manure storage–water courses
Manure storage (structure/ capacity)	180 days	270 days	150 days
Environmental studies requirement	Yes	Yes	No
Mandatory public meetings/ notification	Yes (recommended under guidelines)	Yes	No
Nutrient Management Plans (NMP)	Yes	Yes	No
3rd party review of submissions	By government	By government	—
Who can complete NMP?	Anyone	Farmers	—
Approach to enforcement	Complaint driven	Fines–complaint driven varies	—
Register land for nutrient application	No	Yes for slope of ≥12% (or prohibited by some municipalities)	No
Prohibit winter manure application	No	Indirectly, NMP	Indirectly (waste management)
Limitations on livestock densities/ total size	Indirectly, NMP	NMP	No
Land base requirements for spreading intensive	NMP	Varies by livestock	No (spreading guidelines)
Thresholds defining intensive	300 LU	Varies by livestock	No

[a]Modified from Caldwell & Toombs (2000).
[b]Progress is regularly updated at OMAF website: http://www.gov.on.ca/OMAF/english/agops/index.html
[c]For more information, see Alberta Agriculture, Food and Rural Development (AAFRD) website: http://www.agric.gov.ab.ca/navigation/livestock/cfo/index.html

Table 7.2. Definition of animal unit (AU), livestock unit (LU) or nutrient unit (NU). [a]

Province	Definition of AU, LU, or NU
Manitoba	1 AU = number of animals of a particular type that it takes to excrete 75 kg of total nitrogen in a 12 month period
Alberta	1 AU = number of animals of a particular category of livestock that excrete 73 kg (160 lbs) of total nitrogen in a 12 month period
Saskatchewan	Similar to Alberta definition
Quebec	1 AU = number of animals required to generate 500 kg live body weight
New Brunswick	1 AU = number of livestock needed to produce sufficient manure to meet the nitrogen requirements of 0.4 hectares of hay land
Ontario	1 LU = number of animals required to generate 450 kg live body weight which are housed at one time;
	1 NU = number of animals confined or pastured at one time on a farm unit, that generates enough manure annually to give the fertilizer replacement value of either 95 lbs of nitrogen or 121 lbs of phosphate, whichever is reached first (OMAF 2002)

[a]Modified from OMAF (2000).

In recent years, the legislation and jurisdictions addressing livestock operations across Canada have been actively evolving, probably due to the growing livestock industry and increasing public concern over environmental quality. Table 7.1 provides only the legislation passed before the year 2000. There are two new legislative initiatives passed into law after 2000. In November 2001, the Agricultural Operations Practices Amendment Act was passed by the Alberta legislature which took effect on January 1, 2002 (Brethour et al. 2002). In June 2002, the Nutrient Management Act (Bill 81) was passed in Ontario and will take effect in March 2003.

Manitoba

The hog industry in Manitoba has been growing and expanding in recent years and this has led to construction of large new livestock operations. New policies have been developed at both the provincial and municipal levels in response to this expansion. The siting of large livestock facilities in Manitoba is essentially a municipal responsibility but this has happened with considerable support from the province (Caldwell & Toombs 2000). In some municipalities, conformance to zoning by-laws and a "conditional use permit" are required for the establishment of a new livestock barn. A proposal for the establishment of a new barn may be reviewed by the provincial departments of Environment, Agriculture and Food, Natural Resources, and Rural Development which may recommend changes and impose conditions on municipalities. In March 1998, the Livestock Manure and Mortalities Management Regulation was passed. The regulation is administered under Manitoba's Environmental Act. It prescribes specific requirements for the use, storage and management of livestock wastes, and regulates manure spreading, application rate, management plans, storage structures, composting, spills, transportation, setback distances, as well as the disposal of dead livestock (AGDC 1998). In addition, significant changes from previous years include a ban on spreading of manure to land in the winter and a requirement for a NMP for those operations that exceed 400 AU.

Ontario

On June 27, 2002, new legislation named the Nutrient Management Act (Bill 81) was passed. This legislation will be administered by both the Ontario Ministry of Agriculture and Food (OMAF) and the Ministry of Environment (MOE). The new law will provide province-wide standards for all nutrient management including the application and management of manure, commercial fertilizers and municipal sewage (Brethour et al. 2002). Consultations on the draft regulations have begun since the law was passed. There will be at least 3 stages of consultation, beginning August and October 2002, and the regulations are expected to be in force in March 2003. After the first stage of consultation, two regulations were drafted (OMAF 2002). First, there is now a requirement for a nutrient management plan (NMP) which itself has a large number of requirements. Second, categories of livestock operations are defined that are required to prepare NMPs and when these must be implemented. Categories are based on the size of the operation in terms of nutrient units (NU). All agricultural operations including non-livestock are required to develop an NMP by 2008. Stage 2 and 3 consultations will deal with the issues related to construction and siting of barns and manure storages, land application of manure, setback distances for applying nutrients, restrictions for spreading nutrients on snow-covered or saturated land, livestock access to waterways, manure haulage and transfer, washwater, dead animal disposal, plus other issues. Progress toward finalization of the regulations can be accessed at the OMAF website.

In summary, between and within individual provinces across Canada, different approaches have been developed which reflect regional differences in livestock profiles in the various regions of the country. These differences also reflect the economic role of agriculture, the extent of non-farm development, the community's recent experience with agriculture, the health of the local environment, and the nature and size of the livestock industry (Caldwell & Toombs 2000). The resulting approaches usually develop from regional concerns and these are addressed by a combination of legislation, policy, local by-laws and guidelines concerning manure management.

UNITED STATES

The 1972 Clean Water Act (CWA) included the first set of regulations that affected concentrated animal feedlot operations (CAFO) (Morse 1996). Later, 1987 amendments to the CWA addressed non-point source (NPS) pollution which included pollution arising from agricultural activity (e.g. polluted runoff from fields and feedlots). However, primary responsibility for regulating NPS was essentially given to the states and local agencies to identify pollution problems and develop water quality management programs (Beghin & Metcalfe 1998). In recent years, federal involvement has increased through the activities of the Environmental Protection Agency (EPA) and United States Department of Agriculture (USDA).

Recently, federal rules defined large concentrated animal feedlot operations (CAFO) as point sources of pollution and these are regulated under the National Pollution Discharge Elimination System (NPDES) which issues permits for industrial and municipal wastewater discharge (OMAF 2000). A CAFO is defined as any livestock operation having more than 1000 AU, which is equivalent to 1000 cattle, 2500 swine or 10,000 sheep. Most states have adopted this definition of intensive farming operations. Federal laws have also given individual states the authority to administer a state-specific program relating to "permitting" of agricultural activity likely to cause point source pollution.

The stringency of environmental regulations specific to CAFO varies from state to state. For some of these jurisdictions, regulations have been extensively reviewed in the literature (Metcalfe 2000; OMAF 2000; Beghin & Metcalfe 1998). Table 7.3 shows the approach used by some livestock producing states toward addressing the CAFO. Almost all of these states receive some form of technical and financial support from the state and local governments to address the regulations. As Iowa is the largest hog producing state, its manure and livestock regulations are discussed below.

Iowa

The manure law was passed in 1995 as house file 519 and is essentially administered by the Department of Natural Resources (Kohl & Lorimor 1997). It defined and addressed separation distances between farms and buildings, construction permits, manure management plans, manure application, and drainage tile removal. The law states that all animal feeding operations, regardless of size, cannot be located within 500 ft (152 m) of agriculture drainage wells or their surface intakes or sinkholes, or within 200 ft (61 m) of publicly assessable lakes, streams, or rivers, or 100–1000 ft (30–304 m) from wells. Specific separation distances from private and public buildings are also designated for livestock operations depending on their sizes. Construction permits and manure management plans are required for swine operations with a one time capacity of ≥200,000 lbs live weight. All operations must handle manure so that it will not cause water pollution. Specifically, manure cannot be applied Table 7.3. Legislation regulating concentrated animal feedlot operations (CAFO) in the United States by state. [a] within 200 ft of a sinkhole, cistern, well, agriculture drainage well or its surface intake, or a lake or farm pond, unless it is incorporated in the soil within 24 h or injected. In addition, any drainage tile must be removed ≥50 ft away from earthen manure storage sites or lagoons. It is noted that the

Table 7.3. Legislation regulating concentrated animal feedlot operations (CAFO) in the United States by state.[a]

Jurisdictional context	North Carolina	California	Texas
Intensive farming definition	≥ 100 cattle, 250 swine, 75 horses, 1000 sheep or 20,000 poultry	Follows US EPA definition of CAFO: requirement for NPDES permit if ≥700 AU[b]	Any operation that confines ≥1000 AU for 45 days or more in a 12 month period
State legislation	.0200 Rules	None	Subchapter K
Administered by	Environment and Natural Resources	General water protection rules administered by Water Resources Control Board – regional divisions	Natural Resource Conservation Commission
Status	Enacted in 1993	NPDES permit requirement	In effect in July 1995
Key features of legislation pertaining to CAFO	— Large facilities must register and complete waste management plans — Senate Bill 1217: administration of training and certification program for animal waste managers on swine farms. Only a certified operator can apply animal waste to land — Record keeping is a requirement of Rules — Restriction of development of facilities on floodplain areas	— Threshold limit for nitrates in water is 45 mg/L — Poultry producers must be certified on how to write NMP	— Best management practices (BMP) — Requirement for permit — Air quality authorization — Facility certification — Requirement for training and education programs — Requirement for "pollution prevention plan" (PPP) — Record keeping of manure application — Routine inspection of facilities — Completion of course on animal waste management
Municipal legislation	— Establishment of minimum set-back distances — Zoning	Some counties have stricter regulations than the state	No significant municipal involvement

[a]Modified from OMAF (2000).
[b]1 AU = 1 dairy cow.

Table 7.3. Legislation regulating CAFO in the United States by state (continued).[a]

Jurisdictional context	Nebraska	Minnesota	Pennsylvania
Intensive farming definition	≥ 300 AU[b]	≥ 1000AU[c] or 300 AU that meets at least one of two discharge criteria	Animal density exceeds 2 AEU[d] per acre on an annual basis
State legislation	Livestock Waste Management Act	Rules 7001, 7002, 7020	Nutrient Management Act
Administered by	Environment Quality	Pollution Control Agency	Environmental Protection
Status	Passed in 1998, amended in 1999	Proposed legislation	Enacted in 1993
Key features of legislation pertaining to CAFO	— Construction and operating permits — NMP not mandatory — Annual reports describing NMP strategy required	— Requirement for NPDES permit (existing rules) — Registration of all feedlots with ≥50 AU — Standards for discharge, design, construction, operation, and closure — State Disposal System (SDS) permit (proposed rules) — Implementation of simple pollution control measures — Air quality standard for hydrogen sulphide — Manure management plans for all feedlots with ≥100 AU — Prohibition of manure spreading on frozen or snow-covered soils	— Requirement for a NMP — NMP prepared by a Certified Nutrient Management Specialist — All NMP reviewed by a Public Nutrient Management Specialist — BMP for manure spreading
Municipal legislation	In process of developing zoning ordinances	Increased role and responsibilities in regulations	Limited municipal involvement

[a] Modified from OMAF (2000).
[b] 1 AU = 1 beef animal, 2.5 swine (≥55 lbs), 25 weaned pigs (≤55 lbs), 10 sheep, or 0.7 dairy cow.
[c] 1 AU = 1 dairy cow.
[d] 1 AEU (animal equivalent unit) = 1000 lbs live weight.

legislation established a one-time indemnity fund to help new and existing operations that require construction permits with construction.

EUROPE

Government policies of most European livestock producing countries that address intensive livestock production are predominantly influenced by the European Union's (EU) policies, specifically by the 1991 EC Nitrate Directive. Nitrate policies were first initiated by the EU in the 70s and mainly targeted the quality of drinking water. Today, nitrate levels in groundwater remain the most important element in the environmental policy of many EU countries with respect to agriculture (Wossink & Benson 1999). As the Netherlands and Denmark are the major hog producing countries, their legislative situations specific to hog operations are described below. Summaries of the legislation in both countries are provided in Table 7.4.

The Netherlands

Environmental and legislative aspects of hog production in Holland have been reviewed in several studies (Jongbloed et al. 1999; Beghin & Metcalfe 1998; Wossink & Benson 1999; Neeteson 2000). Environmental regulations affecting the Dutch hog industry include phosphate quotas, waste treatment, storage and spreading, and direct output controls (Beghin & Metcalfe 1998). The regulation of phosphate is based on manure production rights (MPR) quotas. Acceptable practices have been strictly defined for manure spreading in order to minimize nitrogen emissions by leaching or ammonia volatilization. There are restrictions on the time of application (fall and winter) but they vary by soil type and crop. There is an obligation to cover storage facilities used for animal manure (Neeteson 2000). Excess manure on farms must be treated (e.g. incineration, heat treatment, anaerobic digestion or composting) or shipped to other areas. The EU regulates the shipping and export of slurry. The slurry must be processed to meet the microbiological standards defined by EU waste regulations (Table 7.5). There is a well-established industry that treats and ships manure in Holland and markets are available. In addition, each farm is assigned a maximum concentration limit for ammonia which varies by location. Farms in excess can buy ammonia quota from deficit farms within the country. In 1998, a regulation was passed that defined loss standards for nitrogen and phosphate emissions. Farmers are subject to levies when the maximum permissible annual N and P surpluses for farms are exceeded (Neeteson 2000). There is also a Nuisance Act for odour, which requires hog producers to obtain a permit for building farms (Beghin & Metcalfe 1998). The permit limits the maximum number of pigs allowed on the farm and the number is based on the proximity to neighbours. These regulations have adversely affected the supply of hogs for local and international markets in recent years as production has essentially been restricted (Beghin & Metcalfe 1998).

Denmark

Denmark also has similar extensive regulations on manure management and spraying practices regarding nitrogen. Under the 1991 EC Nitrate Directive, the country was identified as a vulnerable zone for nitrate pollution (Beghin & Metcalfe 1998). All acts and regulations since the 80s have been aimed at reducing nitrate pollution from agriculture.

There are many requirements for the establishment and management of hog rearing operations. Hog farms are required to be equipped with manure storage facilities which have to be covered and have a capacity allowing storage of waste up to 12 months. Leaching is prohibited and lagoons are not permitted. All storage facilities have to be lined with concrete. There are also setback distance requirements. Manure storage facilities must be at least 25 m from private water sources (e.g. wells) and 50 m from communal water sources. Hog farmers can spread manure at a rate of <170 kg of nitrogen per hectare per year. There is also a restriction on the spreading time. No spreading of liquid manure is allowed in winter and none is permitted in autumn for solid manure. The manure must be incorporated into the land directly (e.g. injection). Farmers with an excess of manure can spread it on farmland which is mineral-deficit. The transfer from surplus to deficit farms must be documented. Farmers are required to maintain nutrient balance sheets and fertilizer management plans which are later sent to the Ministry of the Environment. Fines are levied on farms that exceed the nitrogen standard. This system has been in place since 1994 (Beghin & Metcalfe 1998). Large hog farms have been discouraged by a recent regulation linking manure and land use (manure/land ratio) for manure disposal. Farmers with 251–500 LU (1 LU = 30 pigs) per year must own 75% of the land required to spread the manure produced while those with ≥500 LU per year must own 100% of the land required, which is equivalent to ≥294 hectares. New farms of this size or larger are no longer permitted (Beghin & Metcalfe 1998).

ECONOMIC IMPACT

Worldwide, environmental concerns linked to livestock production, in particular swine, are increasing. As a result, environmental regulations are becoming more comprehensive in many swine producing countries. The impact of these environmental regulations on the individual livestock producers and the economy of a country as a whole, however, is difficult to assess. This is essentially because environmental regulations differ greatly by province, state or country, and some of these also have local municipalities or counties that adopt zoning and other regulations which affect or limit the establishment or expansion of livestock operations within their jurisdictions (Beghin & Metcalfe 1998).

Table 7.4. Legislation regulating intensive farming operations in the Netherlands and Denmark.[a]

Jurisdictional context	Netherlands	Denmark
Intensive farming definition	$\geq 2LU^b$ /ha	$\geq 2LU^{b,d}$ /ha
State legislation	Numerous acts, main act in force: Act on Manures and Fertilizers	Influenced by EU nitrate Directive
Administered by	Agriculture, Nature Management and Fisheries	Agriculture
Status	Acts in place since 1983	Since 1980
Key features of legislation pertaining to CAFO	- MINAS (Minerals Accounting System): records all minerals (N, P) in and out and remaining on farms. If mineral loss is identified and exceeds the allowable standard, then a levy is applied - Nitrate for drinking water: 25 mg/L^c - Manure injection at spreading, covering storage facilities, building low nutrient emission facilities - No spreading in autumn or winter - Encourage reduction in pork production	- Farmers subject to levy if exceed a quota assigned - Correction of nitrogen losses in animal manure - Test all groundwater against nitrate standards - Reduction of nitrogen emissions to sea: 50% by 2003 - Strict input standard: 170 kg nitrogen/ha - Fertilization plans take account of nitrogen use - Farmers dispose of manure surpluses in the neighbourhood on a contract basis - Manure spreading is prohibited from harvest to February 1 - Farmers must have sufficient manure storage (270 days), and cover liquid manure tanks - Inspection of manure storage every 10 years - 5 year contract if storage capacity is achieved by using facilities on other farms or at biogas plants - Successful establishment of biogas plants
Municipal legislation	Not a large role in regulations. Generally, all EU countries respond to EU requirements which are more strict than those in NA	No significant municipal involvement. As in the Netherlands, Denmark and other EU countries policies are influenced by the EC Nitrate Directive

[a]Modified from OMAF (2000).
[b]1 LU = 1 dairy cow.
[c]50 mg/L of nitrate under EU Nitrate Directive.
[d]$\geq$ 2.3 LU/ha for cattle, $\geq$1.7 LU/ha for pigs.

Table 7.5. Microbiological standards for processed manure in European Commission waste regulations. [a]

Microorganism	Standards
Clostridium perfringens	absent in 1 g of treated product
Salmonella	absent in 25 g of treated product
Enterobacteriaceae	aerobic bacteria count: <1000 CFU/g of treated product

[a]Modified from CEC (2001).

Some studies have attempted to evaluate or even quantify the economic impact that environmental regulations have on livestock production, with varying degrees of success (Innes 2000; Fleming 1999; Beghin & Metcalfe 1998). Fleming (1999) found that using a mathematical model, state requirements in Kentucky did not give enough economic incentives for livestock producers to adopt manure incorporation over surface application as a manure management strategy. Therefore costs of regulation implementation influence manure management strategy. Innes (1999) also developed an empirical model and evaluated the efficiencies of regulatory policies on spatial arrangement of livestock production. However, there were many assumptions made and its practicality for use was not demonstrated.

A review by Beghin & Metcalfe (1998) made some noteworthy evaluations on cost of regulations affecting the competitiveness of hog producing countries in the international market. A discussion using their information follows. The cost of livestock manure management, which differs by farm size and by province or country, influences the competitiveness of the livestock operations locally and globally. In general, large operations meet environmental standards at a lower cost than small traditional operations because expenditures are spread over a larger output. In the EU member countries, especially in Holland and Denmark, concerns about waste disposal have forced producers to adopt costly waste management techniques or to reduce their production capacity. In Holland, the cost of environmental compliance for livestock farms mainly consists of the cost of manure disposal (cost of the actual disposal plus the tax on manure surplus) and the cost of extra waste storage capacity (Wossink & Benson 1999). The total cost of waste disposal and storage for the entire agricultural sector increased from 62 and 49 million guilders, respectively, in 1988 (the first year of manure regulations) to 166 and 156 million guilders, respectively, in 1994. A more stringent set of regulations implemented in 1998 (i.e. MINAS system) will likely cause a further increase in waste storage and disposal costs because it imposes greater penalties. The total cost of environmental regulation (waste handling and treatment, manure production rights, ammonia reduction, building and storage requirements) is estimated between 5–10% of unit cost in Holland. Generally, significant reduction of nitrogen-based emissions to meet the EU nitrate directive is the most costly of the regulations including storage, MPR, accounting, waste disposal. In Denmark, land-use requirements (manure/land ratio) and operation permits have constrained the expansion of hog production.

The establishment of operations having ≥15,000 head is now prohibited. These controls limit export expansion which has directly influenced the industry and economy of the country. Within the U. S., major swine producing states have also been facing increasing costs to meet new environmental standards which vary by state. The costs are increasing at different rates in each state. Unit cost varies greatly across countries, but the share of cost components usually does not. Feed cost usually accounts for 50–60%, labour cost is slightly below 10% in most countries, and capital cost is usually between 15–30% of unit cost. Capital cost is influenced by manure storage requirements which are determined by environmental regulations, but it is still a relatively smaller component compared to the others. Generally, feed and labour costs are relatively lower in North America than in Europe. The labour cost in Denmark is twice as high as that in the U. S. and the feed cost is about 40% larger than that in the U. S. Generally it appears that the environmental compliance cost in the North American hog industry is relatively lower than the cost faced by the EU countries (Beghin & Metcalfe 1998).

CONCLUSIONS

A comparison of the legislative framework and its requirements in a group of hog producing provinces, states and countries is presented in Tables 7.1, 7.3, 7.4. From these data several conclusions were possible (OMAF 2000):

1. It is clear that accounting of nutrient use is more strict in the Netherlands and Denmark (via farm-level nutrient input/output budget) than in Canada and the U. S. (nutrient management plans are used).
2. In all but one jurisdiction (California), livestock and concerns over odour and nutrients from manure were the main reasons for legislation related to ILO. Aside from the federal rules, there is no state legislation in California that specifically addresses ILO.
3. There are an increasing number of regulations that centre on the construction, inspection and monitoring of manure storage systems in order to ensure manure containment and prevent manure leakage. Periodic auditing or random monitoring of NMP is also becoming common in Canada.
4. All jurisdictions strongly discourage the spreading of manure on frozen ground or during the winter season.
5. The definition of AU or LU varies among jurisdictions. The definition may be based on total animal weight, animal weight and manure nitrogen, or even the odour factor.
6. The thresholds for defining an intensive livestock operation vary greatly among jurisdictions. The largest variation is found among Canadian provinces. There is less variation in the US because CAFO is defined by the federal EPA. In the Netherlands and Denmark, the definition is based on land capacity of the operation (animal density) rather than livestock number.

7. The acceptable nitrate levels in groundwater are similar among jurisdictions. In North America, the limit is 45 ppm while in Europe, the limit is 50 ppm.
8. The minimum manure storage capacity required varies among jurisdictions, partially depending on the length of the non-growing season.
9. Generally, European and American producers receive a higher degree of direct financial assistance from government to address environmental issues at the farm level than Canadian producers do.

In addition, none of the jurisdictions has developed any legislation that addresses pathogen contamination of the environment from manure.

REFERENCES AND FURTHER READING

Agricultural Guidelines Development Committee (AGDC). 1998. Farm practices guidelines for hog producers in Manitoba. Manitoba Agriculture, Winnipeg. pp. 4–7, 17–32.

Alberta Agriculture, Food and Rural Development (AAFRD). 1998. Regulatory options for livestock operations: regulation of intensive livestock elsewhere in North America [Online]. URL: http://www.agric.gov.ab.ca/archive/ilo/section4d.html (Accessed: August 27, 2002).

Beghin, J., and M. Metcalfe. 1998. Environmental regulation and competitiveness in the hog industry: an international perspective [Online]. URL: http://www.econ.iastate.edu/research/webpapers/NDN0011.pdf (Accessed: September 5, 2002).

Brethour, C., P. MacGowan, A. Mussell, and H. Mayer. 2002. Proposed new environmental legislation affecting Canadian Agriculture. A special report by the George Morris Centre, Guelph, ON.

Caldwell, W. J., and M. Toombs. 2000. Planning and intensive livestock facilities: Canadian approaches. Revised paper from the Canadian Institute of Planners annual conference, June 1999, Montreal, QC.

Commission of the European Communities (CEC). 2001. Amended proposal for a regulation of the European parliament and of the council: laying down the health rules concerning animal-products not intended for human consumption [Online]. URL: http://europa.eu.int/eur-lex/en/com/pdf/2001/en_501PC0748.pdf (Accessed Sept 29, 2002).

Fleming, R. A. 1999. The economic impact of setback requirements on land application of manure. Land Econ. 75:579–591.

Innes, R. 2000. The economics of livestock waste and its regulation. Am. J. Agric. Econ. 82:97–117.

Jongbloed, A. W., H. D. Poulsen, J. Y. Dourmad, and C. M. C. van der Peet-Schwering. 1999. Environmental and legislative aspects of pig production in The Netherlands, France and Denmark. Livest. Prod. Sci. 58:243–249.

Kohl, K., and J. Lorimor. 1997. Swine manure management and Iowa's manure law [Online]. Iowa State University Extension, Ames, Iowa. URL: http://www.extension.iastate.edu/Publications/PM1700.pdf (Accessed: July 5, 2002).

Metcalfe, M. 2000. State legislation regulating animal manure management. Rev. Agric. Econ. 22:519–532.

Morse, D. 1996. Impact of environmental regulations on cattle production. J. Anim. Sci. 74:3103–3111.

Neeteson, J. J. 2000. Nitrogen and phosphorus management on Dutch dairy farms: legislation and strategies employed to meet the regulations. Biol. Fertil. Soils, 30:566–572.

Ontario Ministry of Agriculture and Food (OMAF). 2000. A review of selected jurisdictions and their approach to regulating intensive farming operations [Online]. URL: http://www.gov.on.ca/OMAFRA/english/agops/otherregs1.htm (Accessed: September 6, 2002).

OMAF. 2002. Nutrient Management Act, September 2002 update. National conference and exhibition of integrated solutions to manure management: working together on challenges and opportunities, September 12 &13, London, ON.

Wossink, A., and G. Benson. 1999. Animal agriculture and the environment: experiences from northern Europe [Online]. Session of Emerging Environmental and Natural Resource Issues in the South, June, Clearwater, Florida. URL: http://www2.ncsu.edu:8010/unity/lockers/users/g/gawossin/papers/sepacf.pdf (Accessed: September 3, 2002).

CHAPTER 8

Risk and Prioritization of Gaps in Knowledge in Managing Pathogens Associated with Intensive Livestock Production

From this review it is clear that the relative risk posed by zoonotic pathogens in hog manure to human health in this region is low, provided guidelines in place for manure handling are followed. Nonetheless, education and vigilance are needed to prevent unforeseen occurrences that might affect the risk level. In contrast, there is significant potential risk to health associated with the application of untreated manure to fields cropped for human food production, particularly fields containing horticultural and other crops likely to be consumed without cooking. Since manure used as fertilizer in Manitoba is continuously commingled while stored in EMS (where it is biologically stable at $\leq 4°C$, the possibility for application of manure containing viable pathogens is high. Risk is minimized by early spring application but ≥ 90 d at $25°C$ are required (we hypothesize) to ensure absence of pathogens from manure. This is an issue of immediate concern that should be addressed by development of minimum hold periods in EMS and/or treatments of manure to shorten pathogen viability. Temperature and pH are two options to consider, but it must be borne in mind that *Campylobacter*, *Yersinia*, and *Listeria* are all alkali-tolerant pathogens. The potential long term risk is that with continued increases in hog production in Manitoba, pathogens will be recycled with greater frequency and at higher concentrations. Eventually, animal densities will be reached in regions of the province that cannot be safely sustained without transport of manure to outlying districts for land application which will further increase pathogen distribution. At present nutrient loading is the only criterion available upon which to base such a management decision and the microbiological threat can only be addressed by default unless work is undertaken to quantify pathogen survival and the accompanying risk to health from the latter source. Due largely to climate, known effective methods for animal waste treatment are not economically practical for the livestock industry in Manitoba at present. However, opportunity may develop in the future as the industry becomes more concentrated, to consider collective or joint ventures to provide for at least secondary treatment of hog manure.

In Taiwan, with a hog density 21 times that of Manitoba, tertiary treatment of all hog waste has eliminated manure handling as a threat to human health. Although in Denmark equally complex systems are used for treatment of most manure (including biogas recovery), there is not the same level of confidence that manure does not contribute to human illness.

A number of gaps in knowledge are identified and they are listed below in the order of importance according to the authors' opinion:

1. There is no one consistent surveillance system to monitor food- and waterborne disease outbreaks and sporadic cases in Canada. National data are usually a compilation of different data sets from provincial governments, the National Enteric Surveillance Program, the Enteric Disease Surveillance System, the Health of Animal Laboratories, the National Laboratory for Enteric Pathogens, and the Bureau of Infectious Disease. To understand the real extent of foodborne disease occurrence in Canada, there is an urgent need for an active, systematic, consistent data collection program at the federal level.

2. There is no statistical database of foodborne illness in Canada which covers the period after 1995 (1996 to present). Because a considerable increase in national pork production occurred after 1996, it is impossible to assess the effect of this increase in production on the health of Canadians. This lack of information may retard policy development important in the prevention of human illnesses in Canada.

3. It is impossible to accurately predict the movement of manure-applied pathogens in Manitoba soil due to the lack of local site-specific studies. It is also unclear whether the setback distances prescribed by provincial guidelines for prevention of nutrient contamination of the environment also help to prevent pathogen contamination following spreading of manure on fields.

4. There is a lack of scientific information on the survival of several important human pathogens in swine manure. These pathogens include *Campylobacter* and *Yersinia enterocolitica*, both of which are frequently excreted by pigs. It is clear that more research is required to close the gap in this area.

5. Presently in Manitoba, the most cost-effect manure treatment system involves the use of earthen manure storage (EMS). Because there is no required minimum length of storage, it is unclear whether manure stored in these structures and exposed to the normal Manitoba climate (long cool months) is sufficiently "treated" to ensure the complete lethality of pathogens. Some pathogens are known to survive in manure for long periods at low temperatures (>one year). Research, specifically field studies

on pathogen survival in an EMS under local climatic conditions should be initiated.

6. Despite the advances in science and technology, there are still many uncertainties associated with methods used for the detection and identification of zoonotic pathogens. Each category of bacteria, viruses and protozoans has its own challenges. Among other factors, while standard methods for each category of pathogen are periodically unreliable, new serological and molecular methods are often complex, expensive, suffer from chemical interference, and require validation for use on environmental samples. There is need for further work to improve sensitivity and specificity of methods for pathogens as well as their ability to determine both viability and virulence.

7. There is no federal regulation governing intensive livestock operations (ILO) in Canada. As the livestock industry is growing and provincial environmental legislation is becoming more comprehensive, it would be beneficial in the longer term if provincial—federal dialogue could be initiated to standardize where possible manure handling regulations—at least the definition of an ILO. Such discussions could improve compliance, provide a vehicle for government assistance and foster a climate in which foreign investment and customer interest are stimulated.

8. Despite a significant number of documented manure-associated disease outbreaks (e.g. Walkerton), there is no jurisdiction in any of the major pork producing countries including Canada that has an environmental regulation specifically addressing prevention of pathogen contamination of the environment from either livestock production or manure application to land. Regulations are based on nutrient loading alone at present.

9. Airborne diseases resulting from agricultural operations are not easily detected, primarily because of our incomplete knowledge of appropriate methods for detection of airborne pathogens. There is a need for more research on airborne disease agents and methods for detection of those pathogens that are associated with agricultural activities.

Despite what we do not know, reviews of this subject have permitted development of some insight which should add to the advantages Manitoba currently enjoys in the global hog industry:

1. The review of literature has identified many documented sources of enteric pathogens and all possible transmission routes available for development of illness in the Canadian population. An important number of waterborne outbreaks and those associated with fresh produce have implicated livestock manure as the source of contamination, which emphasizes the risk of improper manure handling and application. Spreading contaminated manure to land can result in direct contamination of food crops and water sources which are later used for human consumption. In addition, a large portion of food- and waterborne disease occurs as a result of inattention or lack of training of employees involved in food production and handling. These factors were also important with regard

to water use by all groups including producers, processors, distributors, retailers, and consumers.

2. It is clear that storing manure for a period of time before it is applied to land results in a significant reduction in pathogen concentration. It is hypothesized that major bacterial and protozoan pathogens can be eliminated from manure if it is held at 25°C for 3 months. Raw manure should never be spread on frozen fields where there is a potential for spring runoff. It is clear that elimination of pathogens from manure is a critical control point for preventing contamination of food crops and water supplies, and can reduce animal to animal transfer.

3. Using data currently available in Canada, pork does not seem to be a major source of human illness.

4. Some new molecular methods which use or are based on the polymerase chain reaction (PCR) technique can produce more reliable results faster than conventional standard methods. Once problems regarding chemical interference from sample constituents and issues related to target organisms viability are resolved, these methods will prove to be of significant value for pathogen detection in the animal environment.

5. In Canada, livestock manure is not treated as extensively as human waste (wastewater and sludge) to achieve pathogen reduction. Every swine producing country has its own way of handling livestock manure depending on the local needs and conditions (climate, land availability, capital and labour costs, and public concerns). Depending on regional pressures, some countries have adopted more expensive treatment systems (Taiwan, Denmark, and Holland) than others (United States and Canada). Animal density has driven most of these initiatives, however, other factors are involved (human population density, climate). Countries or states with the highest hog densities (expressed in terms of size, hogs/km², relative to Manitoba) include North Carolina, 12.1; Denmark, 13.6; Taiwan, 21.1; and Holland, 38.6, and they have each developed very different systems to deal with manure problems. Hog densities in Quebec, Iowa and Ontario are 5.6, 3.8 and 2.4 times greater, respectively, than in Manitoba at present. An animal density threshold could be used to trigger the need for change in the way manure is handled in Manitoba, however, such a threshold is difficult to establish from the experience of other regions with high levels of hog production. Given the biological stability of hog manure while stored in the Manitoba climate, it is unlikely that the province could sustain the level of hog density currently reported for Quebec. Of other regions considered, the climate in Quebec is most like that in Manitoba.

6. It is clear that microorganisms including pathogens can be transported significant distances and at high concentrations in the environment. Some currently acceptable manure spreading practices can result in pathogen contamination of surface waters which are used as public water sources.

7. Concerns over odour and nutrients from manure are the main reasons that legislation pertaining to intensive livestock production has been developed. Pathogens have not been a major consideration in development of guidelines.

8. There is more strict management of nutrient use from manure in Europe than in North America. It is clear that the problems of nutrient contamination resulting from livestock production have long been established in Europe where livestock production is more intensive.

Index

Note: page numbers in *italics* refer to figures and tables and in **bold** refer to main discussion of the topic